Anna Weinzinger

Linking dynamics of ion channels with biological function

Anna Weinzinger

Linking dynamics of ion channels with biological function

Südwestdeutscher Verlag für Hochschulschriften

Impressum / Imprint
Bibliografische Information der Deutschen Nationalbibliothek: Die Deutsche Nationalbibliothek verzeichnet diese Publikation in der Deutschen Nationalbibliografie; detaillierte bibliografische Daten sind im Internet über http://dnb.d-nb.de abrufbar.
Alle in diesem Buch genannten Marken und Produktnamen unterliegen warenzeichen-, marken- oder patentrechtlichem Schutz bzw. sind Warenzeichen oder eingetragene Warenzeichen der jeweiligen Inhaber. Die Wiedergabe von Marken, Produktnamen, Gebrauchsnamen, Handelsnamen, Warenbezeichnungen u.s.w. in diesem Werk berechtigt auch ohne besondere Kennzeichnung nicht zu der Annahme, dass solche Namen im Sinne der Warenzeichen- und Markenschutzgesetzgebung als frei zu betrachten wären und daher von jedermann benutzt werden dürften.

Bibliographic information published by the Deutsche Nationalbibliothek: The Deutsche Nationalbibliothek lists this publication in the Deutsche Nationalbibliografie; detailed bibliographic data are available in the Internet at http://dnb.d-nb.de.
Any brand names and product names mentioned in this book are subject to trademark, brand or patent protection and are trademarks or registered trademarks of their respective holders. The use of brand names, product names, common names, trade names, product descriptions etc. even without a particular marking in this work is in no way to be construed to mean that such names may be regarded as unrestricted in respect of trademark and brand protection legislation and could thus be used by anyone.

Coverbild / Cover image: www.ingimage.com

Verlag / Publisher:
Südwestdeutscher Verlag für Hochschulschriften
ist ein Imprint der / is a trademark of
OmniScriptum GmbH & Co. KG
Heinrich-Böcking-Str. 6-8, 66121 Saarbrücken, Deutschland / Germany
Email: info@svh-verlag.de

Herstellung: siehe letzte Seite /
Printed at: see last page
ISBN: 978-3-8381-5095-6

Zugl. / Approved by: Wien, Universität Wien, Habil., 2015

Copyright © 2015 OmniScriptum GmbH & Co. KG
Alle Rechte vorbehalten. / All rights reserved. Saarbrücken 2015

"Everything that living things do can be understood in terms of the jiggling and wiggling of atoms"

Richard Feynman

For my family

TABLE OF CONTENTS

1 4
1   WHY STUDY ION CHANNELS? 7
1.1   Brief history of selected ion channel research milestones 8
1.2   Conserved architecture of voltage gated ion channels 9
1.3   Ion binding and selectivity 13
1.3.1 K^+ selective channels 13
1.3.2 Na^+ and Ca^{2+} selective channels 14
1.4   Gating mechanisms of cation channels 16
1.4.1   Inactivation gating 17
1.4.2   Allosteric coupling between selectivity filter gate and inner pore gate 18
2   PHARMACOLOGY 19
2.1   The hERG K^+ channel 21
3   REFERENCES 24
4   SHORT SUMMARIES OF PAPERS DESCRIBED IN THIS HABILITATION THESIS 32
4.1   Timothy Mutation Disrupts the Link between Activation and Inactivation in Cav1.2 Protein . 33
4.2   Toward a Consensus Model of the hERG Potassium channel 34
4.3   Different inward and outward conduction mechanisms in NavMs suggested by molecular dynamics simulations 36
4.4   Probing the Energy Landscape of Activation Gating of the Bacterial Potassium Channel KcsA 37
4.5   In silico analysis of conformational changes induced by mutation of aromatic binding residues: consequences for drug binding in the hERG K^+ channel 38
4.6   Structural insights into trapping and dissociation of small molecules in K^+ channels 39
4.7   Molecular determinants for activation of human ether-à-go-go-related Gene 1 Potassium channels by 3-nitro-N-(4-phenoxyphenyl) Benzamide 40
4.8   ICA-105574 interacts with a common binding site to elecit opposite effects on inactivation gating of EAG and ERG Potassium channels 41
5   PAPER REPRINTS 42
5.1   Paper #1 43
5.2   Paper #2 52
5.3   Paper #3 66
5.4   Paper #4 80
5.5   Paper #5 90
5.6   Paper #6 100
5.7   Paper #7 112
5.8   Paper #8 121

Significance:

Ion channels are transmembrane proteins, regulating a multitude of biological processes by controlling passive flow of ions across cell membranes. They play a key role in diverse physiological processes such as signal perception, neuronal communication, regulation of blood pressure, and cell proliferation. With more than 60 ion channel diseases identified to date, ion channels have great potential for the treatment of human disease. Drugs that modulate ion channels are in clinical use for the treatment of cardiac arrhythmias, epilepsy, diabetes or hypertension. Owing to the complexity of these channels and the lack of 3D structures for most family members, rational drug design is very challenging. For a large number of family members no specific ligands have been identified to date. Another complicating factor for drug development arises from the fact that these channels undergo large conformational transitions during gating, resulting in different binding affinities for different channel states and changes in binding site accessibilities. Thus, a better understanding of the structure, gating dynamics and molecular mechanisms underlying disease is urgently needed.

My research addresses several of these fundamental questions. Specifically, papers #1 and #2 describe strategies and methods to develop accurate and reliable 3D structure models of ion channels for which no atomic resolution structures are available. Papers #3, #4 and #6, provide a detailed characterization of gating mechanisms and dynamics guided by bacterial ion channel x-ray structures and finally selected examples of drug – receptor interaction analyses – in relation to dynamics are presented in papers #5 - #8.

Highlights:

- Development of a validated hERG K^+ channel pore model useful for investigating drug binding in detail
- Identification of a structure motif in Ca_v channels, linked to an inherited ion channel disease, essential for normal gating
- Newly identified binding site for the hERG activator ICA105774
- Novel insights into drug trapping and drug induced changes in gating dynamics
- Novel insights into conductance mechanisms in bacterial sodium channels, providing insights into early inactivation steps

Papers by the author discussed in this habilitation thesis

#1 Timothy mutation disrupts the link between activation and inactivation in Cav1.2 protein. Katrin Depil[#], Stanislav Beyl[#], **Anna Stary-Weinzinger**[#], Annette Hohaus, Eugen Timin, Steffen Hering *Journal of Biological Chemistry*, 2011, 286,36, 31557-31564 [#]equal contribution

#2 Toward a consensus model of the hERG potassium channel. **Anna Stary***, Sören J. Wacker, Lars Boukharta, Ulrich Zachariae, Yasmin Karimi-Nejad, Johan Åqvist, Gert Vriend, Bert L. de Groot, *ChemMedChem 2010*, 5, 455-467

#3 Different inward and outward conduction mechanisms in NavMs suggested by molecular dynamics simulations. Song Ke, Eugen Timin, **Anna Stary-Weinzinger*** *PLOS Computational Biology*, 2014, 10, 7, e1003746

#4 Probing the energy landscape of activation gating of the bacterial potassium channel KcsA. Tobias Linder, Bert L. de Groot, **Anna Stary-Weinzinger*** *PLOS Computational Biology*, 2013, 9, 5, e1003058

#5 In silico analyses of conformational changes induced by mutation of aromatic binding residues: consequences for drug binding in the hERG K^+ channel. Kirsten Knape, Tobias Linder, Peter Wolschann, Anton Beyer, **Anna Stary-Weinzinger*** *PLOS ONE*, 2011, 6, 12, e28778

#6 Structural insights into trapping and dissociation of small molecules in K^+ channels. Tobias Linder, Priyanka Saxena, Eugen Timin, Steffen Hering, **Anna Stary-Weinzinger*** *Journal of Chemical Information and Modelling*, 2014, 54(11):3218-28.

#7 Molecular determinants for activation of the human ether-à-go-go-related gene 1 potassium channels by 3-nitro-N-(4-phenoxyphenyl) benzamide. Vivek Garg, **Anna Stary-Weinzinger**, Frank Sachse, Michael C. Sanguinetti *Molecular Pharmacology*, 2011, 80, 4, 630-637

#8 ICA-105574 interacts with a common binding site to elicit opposite effects on inactivation gating of eag and erg potassium channels. Vivek Garg, **Anna Stary-Weinzinger**, Michael C. Sanguinetti *Molecular Pharmacology*, 2013, 83, 805-813

*corresponding author

1 Why study ion channels?

Ion channels are a large, diverse group of pore-forming membrane proteins that enable selective ion conductance across otherwise ion impermeable lipid membranes of cells. These channels are found in the plasma membrane of nearly all cells and many intracellular organelles. Ion channels play a fundamental role in numerous physiological processes including the generation of action potentials in electrically excitable cells, regulation of cell volume and muscle contraction to name but a few.

Ion channels are linked to a multitude of human diseases (referred to as channelopathies) including cardiac arrhythmias, epilepsy, pain, diabetes and hypertension (Kim, 2014; Moreau et al., 2014; Olson and Terzic, 2010). With more than 60 channelopathies identified to date (Bagal et al., 2013), these proteins represent an important target class for drug discovery. Further, recent studies indicate that ion channels play a central role in cancer (for recent reviews see Pardo and Stühmer, 2014; Huang and Jan, 2014; Fraser et al., 2014; Chen et al., 2014).

Despite, their role in human disease, less than 20% of ion channel targets are commercially exploited today (Overington et al., 2006), owing to the remarkable complexity of these proteins. **Thus, the need to understand the molecular mechanisms of ion channels in health and disease remains exceptionally high**.

My research over the last years is primarily focused on cation channels. In particular my research interests lie in the investigation of structure, dynamics and ligand interactions of voltage gated potassium (K_v), sodium (Na_v) and calcium (Ca_v) channels as well as ligand gated inwardly rectifying potassium (K_{IR}) channels using computational methods. In this habilitation thesis only research projects concerning K_v, Na_v, and Ca_v channels are presented.

In the following chapters the basic functional concepts and architecture of voltage gated ion channels (VGICs) are described in detail. Excellent reviews about other ion channels such as anion channels and ligand gated ion channels can be found in the literature (Lemoine et al., 2012; Stölting et al., 2014; Colombini, 2012). Details about the computational methods used in **papers #1 - #8** can be found in the respective methods sections of the publications.

1.1 Brief history of selected ion channel research milestones

Ground-breaking insights into the mechanisms how voltage-gated ion channels give rise to propagating action potentials are based on the pioneering work by Hodgkin, Huxley and Cole, in the early 50ies, who published their theories of action potentials in squid giant axons (Hodgkin and Huxley, 1952; Hodgkin and Huxley, 1952b, Cole, 1979). Another, very crucial invention for investigating ion channels concerns the development of the patch-clamp electrophysiology method by Neher and Sakmann in 1976 (Neher and Sakmann, 1976). This state-of the art method enables the analysis of the functional properties of ion channels in great detail. Another milestone in the field followed in 1982, when the first ion channel (α-subunit of the acetylcholine receptor) was successfully cloned (Noda et al., 1982). Since then, molecular biology methods enabled the discovery of a plethora of ion channels. As of today, more than 400 ion channel members are known, accounting for ~ 1,5% of the human genome (http://www.guidetopharmacology.org; Pawson et al., 2014). Another breakthrough came in 1998 with the first crystal structure of a bacterial K^+ channel solved at atomic level in the MacKinnon lab (Doyle et al., 1998). For a more complete review of ion channel research milestones see for example Catterall, 2012.

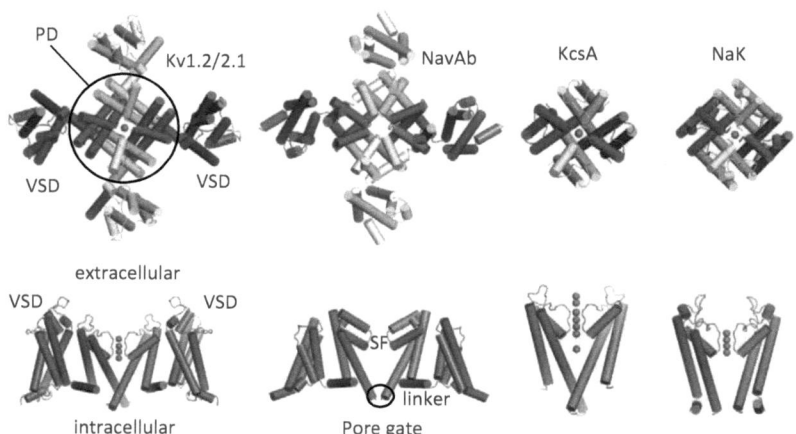

Figure 1; upper panel: top view of selected ion channels; 2 opposite subunits are shown in dark grey and light grey, respectively; K^+ ions are shown as spheres; lower panel: side view of the respective channels.

1.2 Conserved architecture of voltage gated ion channels

Since then, tremendous advances in our understanding of the atomic structure of ion channels were enabled by the successful determination of crystal structures of different members of the K$^+$ channel family (Zhou et al., 2001; Uysal et al., 2009; Cuello et al., 2010; Miller and Long, 2012; Brohawn et al., 2012; Jiang et al., 2003; Long et al., 2005; Tao et al., 2009). Very recently the first bacterial Na$^+$ channel structures were crystallized (Payandeh et al., 2011).

Several structures of other bacterial sodium channels have followed since (Zhang et al., 2012; Tsai et al., 2013; Shaya et al., 2014) shedding light on fundamental relationships within the voltage gated ion channel superfamily. The co-crystallization of the bacterial Na$^+$ channel Na$_v$Ms with brominated Na$^+$ channel blocker compounds by the Wallace lab enabled a first glimpse into small molecule interaction with Na$_v$ channels (Bagnéris et al., 2014). These studies provide tremendous opportunities for the ion channel research field and hold great promise for a better understanding of ion channel function in health and disease. Further these studies suggest that bacterial structure information will be useful to study the pharmacologically interesting mammalian channels, using structure information from bacterial channels.

A complete list of currently available high resolution structures of ion channels can be found in the "membrane proteins of known structure" database under the following link: http://blanco.biomol.uci.edu/mpstruc/. Remarkably, despite often very low sequence identity (e.g. <20 % between the pore domains of KcsA and NaK), recent x-ray crystallography success revealed that these channels share a conserved architecture (Fig. 1).

The architecture of most members comprises a voltage sensing domain (VSD) and a pore domain (PD). In case of K$_v$ channels, four subunits form a homotetrameric channel, in case of Na$_v$ and Ca$_v$ channels the pore forming subunits are composed of a single polypeptide chain folding into four homologous transmembrane domains. Transmembrane helices S1 to S4 form the VSD, whereas transmembrane helices S5 (M1) and S6 (M2) form the PD that includes the selectivity filter (SF) region.

As expected, structural data revealed how the different channel subtypes are fine-tuned for different functions. In particular, the selectivity filter motif is different between channels with different ion selectivity. As described in more detail in chapter 1.3.1, the selectivity filter of K_v and Na_v channels is composed of several residues lining a narrow region close to the extracellular side. In both channel types the two innermost positions are lined by backbone carbonyl oxygen atoms. In contrast, the residues above (closer to the extracellular side) differ in their geometry. While in K^+ channels the upper part of the SF is also lined by backbone carbonyl atoms, in Na_v channels the side chains of selected residues (in case of bacterial Na_v channels: EEEE, in case

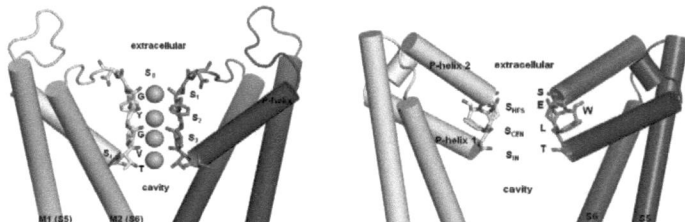

Figure 2 Comparison of KcsA (left side) and Na_vAb (right side) x-ray structures the filter sequence in both channels is shown in stick representation; for clarity, only two opposing subunits are shown. K^+ ions are shown as spheres.

of eukaryotic channels: DEKA) are oriented towards the lumen and play a crucial role for selective conduction (see Figure 2). Another well conserved segment concerns the P-helix, which stabilizes the filter region. The major difference between K^+ and Na^+ channels in this region concerns the existence of a second P-helical segment in the latter subfamily, which is probably needed to stabilize the extended (widened) geometry of the filter as shown in Figure 2.

Another interesting feature revealed in ion channel structures concerns the size of lateral pore openings termed fenestrations (see Figure 3), which are suggested to play an important role for lipid modulation and might provide access pathways for lipophilic drug molecules in some channels (Raju et al., 2013; Martin and Corry, 2014; Boiteux et al., 2014a). While these lateral openings are seemingly large enough for small molecule access in Na_v channels, the size of these openings, formed between neighbouring S6 segments, varies considerably in the K^+ channel family. For example these openings are rather narrow in KcsA (Figure 3, right side) but quite large in recently crystallized two-pore K^+ channels (Brohawn et al., 2012). Despite their size, it is not clear if these fenestrations serve as access routes for lipophilic drugs, as suggested in Na_v channels (Rapedius et al., 2012). From x-ray crystallography we know that the size of the fenestrations varies depending on the gating state. Interestingly, it was recently suggested that the size might vary during inactivation (Payandeh et al., 2012). The positioning of the fenestrations coincides with the newly identified binding site for the small molecule activator ICA105574 in hERG K^+ channels, which significantly slows inactivation (see paper #7 for details and chapter 1.5.1). Unfortunately, due to the lack of x-ray structures in hERG, no details about the size of lateral openings in these channels are known. An interesting hypothesis about the fenestrations in hERG was made by Zachariae et al (Zachariae et al., 2009), who suggested that this region might explain, why even very large molecules can bind to the hERG cavity. This group proposed that lateral openings in hERG might facilitate block of bulkier molecules, which could protrude towards the outer pore helix (S5) through these lateral openings.

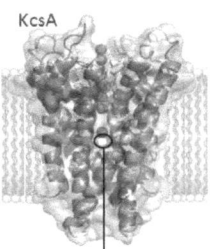

Figure 3 – Fenestration site in KcsA.

Probably, the most important differences with respect to function in the architecture of ion channels can be found in the extra- and intracellular domains. This is of special interest, since more and more studies are emerging, suggesting the important role of these regions as interacting sites with accessory subunits and other regulatory proteins. Thus, it is very likely that these channel regions constitute major regulatory domains. For a recent review see Lee et al., 2014.

The overall similar architecture, particularly in the transmembrane domains, provides a tremendous advantage for comparative structure modelling approaches. Despite recent success in x-ray crystallography and more recently in cryo-EM (Liao et al., 2013; Efremov et al., 2014) to obtain high resolution structures of selected ion channel members, computational structure prediction methods are urgently needed to fill the gap between available sequence information and structure knowledge. In particular, the large number of ion channel members necessities the use of other methods to derive at a more complete picture of the structural landscape of this large protein family.

In **papers #1** and **#2**, two examples of **comparative structure modelling** strategies are described, aimed at obtaining **reliable, high quality structure models of ion channel family members**, for which structure information is missing so far.

The quality of homology models does not only depend on the computational methods employed and the careful evaluation by theoretical and experimental means, but is influenced a great deal by the quality and similarity of available 3D structure templates. In **paper #1** modelling of the pore domain of an important member of the Ca_v channel family, the $Ca_v1.2$ channel, is presented (Depil et al., 2011). The challenge in this subfamily lies in the fact that to this date no atomic resolution structure of any members of this family is available. Thus modelling necessities the use of structure templates from more distantly related family members such as K_v or Na_v channels. At the time of this research project (2010 - 2011) only structures from K^+ channel family members where available, rendering modelling strategies exceptionally challenging. Nevertheless, to obtain experimentally testable hypothesis of this functionally important ion channel class, a modelling approach was attempted in close collaboration with electrophysiology experiments. A particular aim of this paper was to provide first structural insights into a rare $Ca_v1.2$ channelopathy termed Timothy Syndrome (TS).

TS is an autosomal dominant disease characterized by de novo missense mutations (most frequently by point mutations G406R and G402S) in the CACNA1C gene, which encodes the $Ca_v1.2$ channel. Very recently a novel mutation at position I1166T was identified, leading to TS (Boczek et al., 2014). This severe disease is characterized by multi-organ dysfunction, including lethal arrhythmias, congenital heart disease, webbing of fingers and toes, immune deficiency, hypoglycemia, cognitive abnormalities, seizures, mental retardation and autism (Splawski et al., 2004). Unfortunately, management of TS is challenging and prognosis is poor. According to Splawski et al (Splawski et al., 2006) the average age of death is 2.5 years and no drugs are currently available to cure this rare disease. Treatment of TS includes use of β-blockers to maintain QT interval stability in order to prevent ventricular tachyarrhythmia. Recently, a novel promising pharmacotherapeutical approach was reported in a female Chinese TS patient. In this study it was shown that mexiletine inhibits late Na^+ currents ($I_{Na,L}$), thereby significantly reducing the risk of arrhythmic in this TS patient (Gao et al., 2013), raising hopes that pharmacological intervention might be feasible for patients with TS.

Our research, as described in **paper #1**, (Depil et al., 2011), contributes to the investigation of the biophysical properties of mutation G402S and possible structural consequences on the $Ca_v1.2$ channel structure. Multi-sequence analyses combined with comparative structure modelling revealed that the TS mutation, located close to the gate of $Ca_v1.2$, is not only highly conserved in nearly all Ca_v channel family members, but is part of a unique structure motif, (which we termed G/A/G/A), essential for normal channel gating. The functional uniqueness of this structure motif was recently confirmed by *a posteriori* experimental and mathematical investigations by Beyl et al (Beyl et al., 2012; Beyl et al., 2014).

In **paper #2** (Stary et al., 2010) a modelling and validation strategy for the hERG K^+ channel, one of the most challenging ion channels for drug development, due to its "off-target" pharmacology, is presented (for details about hERG see chapter 1.5.1).

The availability of atomistic structure models from experimental methods (x-ray, cryo-EM, NMR) and theoretical modelling approaches (e.g. comparative modelling) enables to address many fundamental questions concerning the biological function of ion channels. One of the most intriguing questions that captivated generations of scientists involves the elucidation of the **selective conduction** of these channel pores **at a rate close to diffusion limit**.

1.3 Ion binding and selectivity

Elucidating the molecular mechanisms enabling extremely rapid (up to 10^8 ions per second), highly selective ion flux (e.g. K^+ over Na^+ factor ~ 10,000:1; Hille, 2001) represents one of the most fundamental issues ion channel research. Classical biophysical evidence for aqueous pores for ion diffusion (Hille, 2001) were confirmed by several x-ray structures of K^+ and Na_v^+ channels.

1.3.1 K^+ selective channels

Common to all K^+ channels seems a narrow selectivity filter with a signature sequence TVGYG, which enables exquisite selectivity and conductance rates near diffusion limit. The best understood mechanisms concerning selective conductance are available for K^+ channels. Structure information from several K^+ channels revealed that the selectivity filter is structurally conserved. As shown in Fig. 2 (left side), four channel subunits contribute to a linearly extended backbone of carbonyl oxygen atoms aligned toward the center of a four-fold symmetric filter pore. The carbonyl oxygen atoms together with the side-chain oxygen atoms of threonine form a series of connecting binding sites, termed $S_1 - S_4$ (Aqvist and Luzhkov, 2000). Additionally, binding sites at the extracellular entrance (S_0) and in the cavity (S_{cav}) have been observed (Zhou et al., 2001). From x-ray structures it was interpreted that two consecutive K^+ ions are alternated with water molecules. Based on this structural information, a widely accepted hypothesis suggesting co-translocation of ions with water molecules in a coupled knock-on fashion was proposed (for recent review see Furini and Domene, 2013). Intriguingly, a very recent study by Köpfer et al (Köpfer et al., 2014), suggests that in contrast to this previous widely accepted hypothesis a direct Coulomb knock-on mechanism might be a more suitable mechanism to explain the highly efficient K^+ conductance observed experimentally. Their hypothesis is based on extensive molecular dynamics simulations of several K^+ channel structures with applied transmembrane ion gradients (Kutzner et al., 2011), and supported by a careful reanalysis of the previously published K^+ crystal structures. Importantly, this new hypothesis provides a good explanation, why linear conductance rates are observed experimentally over a wide range of K^+ concentrations (Morais-Cabral et al., 2001).

1.3.2 Na^+ and Ca^{2+} selective channels

Much less is known about the conduction and selectivity mechanism of Na_v and Ca_v channels. With the recent success in crystallizing bacterial homologs of Na_v channels (Payandeh et al., 2011; Zhang et al., 2012; McCusker et al., 2012) a structural interpretation analogous to K^+ channels becomes possible. Based on the first x-ray structure of Na_vAb (Payandeh et al., 2011), three ion binding sites in the SF were proposed: an electronegative region close to the extracellular side, formed by the side chains of four glutamic acids (E177, EEEE motif) termed S_{HFS}, together with two rings of carbonyl oxygen atoms contributed by leucine and threonine residues, which are denoted sites S_{CEN} and S_{IN} (see Figure 3, right side). Since the filter region in Na_v channels is wider compared to K^+ channels, ions might be coordinated by water molecules. This hypothesis is in line with recent MD simulations (e.g. Carnevale et al., 2011; Corry and Thomas, 2012; Ke et al., 2013). Equilibrium MD simulations together with free energy calculations confirm the Na^+ binding sites (Furini and Domene, 2012; Zhang et al., 2013; Stock et al., 2013; Ke et al., 2013, **paper #3** 2014) suggested by the Catterall lab based on the Na$_V$Ab crystal structure. Further, MD simulations suggest that double occupancy of Na$^+$ ions at the same site (sites S_{HFS} and S_{CEN}) form stable configurations (Corry and Thomas, 2012; Stock et al., 2013). Another important difference to K$^+$ channels concerns the stability of the SF. While in K$^+$ channels, K$^+$ ions are essential for stabilizing the selectivity filter (Cuello et al., 2010), the filter of the Na$_V$ channel remains stable, even in the absence of ions in the filter region and might conduct water in this "ion-free" state (Carnevale et al., 2011). Recent studies indicate that despite different filter stabilities in the "ion-free" state, K$^+$ channels can conduct water in this "collapsed" configuration as well (Hoomann et al., 2013).

We have recently shown (**paper #3**, Ke et al., 2014) that Na$^+$ in- and efflux occur by distinct mechanisms. In particular, ion efflux is two-fold lower compared to influx, due to the asymmetric structure of the SF, with respect to directionality. Our simulations revealed an energy barrier for sodium efflux, due to the "more hydrophobic" nature at a new site termed S_{BAR} (formed by the β-ch2 and γ-ch2 groups of the E side chain). Interestingly, conformational changes at the site of the glutamic acids were observed, which significantly influence the height of the energy barrier at this site. In particular, our simulations revealed that flipping of a single E side chain resulted in lowering of the energy barrier, while flipping of two or more E side chains resulted in a complete stop of ion efflux. This is particularly interesting, since bacterial Na$_V$ channels do not contain an inactivation particle (N-type inactivation), stopping ion flux during prolonged membrane depolarization as suggested for mammalian channels. It is assumed that these channels only possess C-type inactivation to control/stop efflux (Irie et al., 2010). It is possible that the conformational changes at the glutamic acid side (EEEE locus) might be a first step towards initiating inactivation gating in these channels. Recent evidence suggests however that the cytoplasmic domain might also contribute to inactivation (Bagnéris et al., 2013) in bacterial channels. Thus, clearly further studies are needed for a better understanding of inactivation mechanisms in these channels.

It remains to be investigated, if studies aimed at understanding conduction, selectivity and filter inactivation in bacterial Na$_V$ channels, are able to provide insights into these functions in mammalian channels, since the selectivity filter signature sequence (EEEE) is different (DEKA in mammalian channels). Recently, a first in silico mutational attempt to investigate the role of the

DEKA motif in ion discrimination and conductance in mammalian channels, based on mutating the EEEE locus to DEKA in the bacterial Na_vRh channel was described (Xia et al., 2013).

Another only partly understood fundamental question concerns the structural basis of the gating mechanisms (activation, deactivation, and inactivation) of ion channels.

1.4 Gating mechanisms of cation channels

Ion channels control the passage of ions across cell membranes via actively changing their conformation from open (i.e. conductive) to closed (i.e. nonconductive) states. The kinetics of these mechanisms can be very rapid, on the order of tens of microseconds to a few milliseconds (Hille, 2001). Structural studies, particular on bacterial potassium channels, have provided tremendous insights into the general rearrangements occurring upon activation gating (i.e. channel opening). From x-ray structure of KcsA and a series of experimental investigations including electron paramagnetic resonance spectroscopy it can be deduced that the lower parts of the inner helices move during gating to widen and narrow a constriction site formed by the bundle crossing of these inner helices (Kelly and Gross, 2003; Perozo et al., 1999; Shimizu et al., 2008; Zimmer et al., 2006). To obtain insights into the transition pathways between conformational states we and others have performed computational studies (e.g. Linder et al., 2013, **paper #4;** Biggin and Sansom, 2002; Denning and Woolf, 2010; Shrivastava and Bahar, 2006; Tikhonov and Zhorov, 2004). The advantage of our study is that we did not have to compare structures from different K^+ channel family members or had to rely on homology models for missing gating states, as was the case in most other previous computational gating studies. Due to the availability of atomic resolution structures of different gating states of KcsA at the time of our study, our simulations provided detailed insights into dynamical side chain rearrangements of gating not known previously. Further, the underlying accurate structure information enabled the calculation of the underlying energetic landscape of pore gating, thus quantifying the importance of local and global structural changes during activation gating in KcsA. A detailed summary of this paper can be found in chapter 4.3 on page 36 and in the paper reprint.

A question, which cannot be solved using "pore only" bacterial K⁺ channel structures such as KcsA, concerns the mechanism how ion channels can open and close in response to changes in membrane voltage.

When the membrane is depolarized the voltage sensing domains (VSD) of ion channels undergo conformational changes from a so called "resting state" towards an activated state. These VSD movements, associated with the transfer of electric charges are transferred to the pore domain, which opens in response to these movements, thereby allowing selective ion conduction. A scheme about the full gating cycle as seen in a recent MD simulation can be found in Jensen et al, 2012. Since atomic resolution structures of voltage gated ion channels are only available with the VSD domain in the active conformation one of the most fundamental questions over the last decades concerned questions about the structure(s) of the resting state, how the VSD domain moves in response to membrane depolarization and how this leads to pore opening. To address these questions, numerous experimental and computational investigations have been performed. From these studies a consensus model, how voltage dependent gating in voltage gated ion channels might occur emerged (for a detailed summary see, Vargas et al., 2012). Briefly, these studies suggest the existence of different resting states in agreement with experimental data (e.g. DeCaen et al., 2009; Henrion et al., 2012; Lacroix et al., 2012). Compared to the open state x-ray structures the S4 segment is vertically displaced and all gating charge residues (arginine and lysine residues) form favourable interactions, either with negatively charged residues from segments S1 to S3 or interact with polar head groups from the lipid bilayer. According to modelling studies, voltage-dependent activation occurs via a so called helical screw-sliding mechanism, which means that the S4 segment moves along its principal axis and retains its helical conformation, while sequentially forming salt bridges with residues from other VSD helices (S1 – S3). The first proposals of such a gating model came independently from the labs of Catterall and Guy (Catterall, 1986; Guy and Seetharamulu, 1986).

1.4.1 Inactivation gating

Fast and slow inactivation mechanisms are important to terminate ionic conductance. Fast inactivation usually operates on the millisecond timescale and has been observed in all types of voltage gated cation channels. Experimental studies led to the hypothesis that fast inactivation occurs via a "hinged-lid" mechanism, whereby a tethered cytoplasmic inactivation gate occludes the pore. In contrast, slow inactivation is assumed to involve structural rearrangements at the selectivity filter and involves an allosteric crosstalk with the activation gate (for review see Kurata and Fedida, 2006). The physiological importance of inactivation is highlighted by disease mutations such as hyperkalemic periodic paralysis (Cannon, 2007) or neurological and psychiatric disorders (Adelman et al., 1995). Understanding the structural basis of inactivation is further highly important for drug development, since clinically relevant ion channel blocking drugs can bind to slow or C-type inactivated states and/or stabilize inactivated channel states (e.g. Wang et al., 2011).

The structural mechanisms of C-type inactivation are best understood in the bacterial K^+ channel KcsA (Cuello et al., 2010). In combination with detailed mutagenesis and electrophysiology studies (expertly summarized in Hoshi and Armstrong, 2013) as well as computational investigations structural hypotheses emerged.

1.4.2 Allosteric coupling between selectivity filter gate and inner pore gate

Analysis of multiple x-ray structures of KcsA channels in open and partially open conformations in combination with computational investigations revealed a mechanism how activation gate movements might bias conformational changes in the selectivity filter towards a nonconductive inactivated state. In particular, an aromatic residue in S6, located close to the selectivity filter (F103), seems to be very important in this respect (Pan et al., 2011). A similar crosstalk between SF and activation gate has also been postulated for Shaker K^+ channels (Peters et al., 2013). Interestingly, these coupling effects between F103 and the S6 gate could also be observed in our pore gating simulations of KcsA (**paper #4**, Linder et al., 2013). An important difference to the study by Pan et al (Pan et al., 2011), however, lies in the fact that our simulations reveal that while conformational changes at position F103 are important for activation gating, they do not necessarily enforce the filter to collapse and inactivate. In all of our ten ED simulations the filter was proposed to remain in a conductive state, which was confirmed *a posteriori* by Köpfer et al (Köpfer et al., 2014). A recent study by Ostmeyer et al (Ostmeyer et al., 2013) highlights the importance of water molecules behind the selectivity filter for inactivation and slow recovery in KcsA.

C-type inactivation mechanisms are also of particular interest for the hERG K^+ channel research community, since there is compelling evidence that inactivation influences drug affinity (reviewed in Vandenberg et al., 2012). How small molecule interactions in hERG and hEAG K^+ channels can contribute to a better understanding of inactivation in these ion channels is described in **papers #7** and **#8** (chapter 1.5.1).

Very recently, a set of bacterial Na^+ channel structures in putatively inactivated states was published by the Catterall lab (Payandeh, 2012). Since bacterial Na_v channels do not contain an inactivation particle, these structures are assumed to provide first structural insights into C-type (slow) inactivation in Na_v channels. Interestingly, these new x-ray structures reveal asymmetric rearrangements of the S6 helices, as well as changes in the selectivity filter region, possibly induced by structural rearrangements in the activation gate. Thus, these structures can provide first insights into the allosteric coupling between SF and S6 gates. A clear difference to the structural data obtained in K_v channels is the fact that in K^+ channel x-ray structures inactivation was accompanied by an open pore gate, while the new Na_v channel structures reveal a closed inactivated state. It remains to be investigated how similar or dissimilar structural rearrangements during slow or C-type inactivation are between different channel types. Clearly, due to the differences in SF structure and stability (see chapter 1.3.2), the fine-tuned structural rearrangements might be different – however as implied from functional studies the allosteric cross-talk between channel segments seems to be evolutionary conserved. Interestingly, a cross talk between SF and gate has recently been suggested for a non-voltage dependent inward rectifier K^+ channel (Clarke et al., 2010). However, this hypothesis is still under discussion (Bavro et al., 2012).

2 Pharmacology

Due to the ubiquitous expression of ion channels in all tissues and cell types of the human body and their modulation by disease it is clear that **ion channels represent a very important class of drug targets**. Table 1 lists examples of currently marketed ion channel drugs.

Historically, ion channel modulator discovery occurred by chance. For example, the natural product cocaine extracted from coca leaves was used in clinics in the 1880s for its analgesic properties. Due the toxic side effects in the CNS and cardiovascular system, medicinal chemists synthesized new derivatives, leading to the development of local anaesthetics, such as lidocaine. However, it was not until the late 1950's that these compounds where shown to be selective sodium channel blockers (Cox and Gosling, 2014). With the tremendous knowledge accumulation in the ion channel field and particularly the possibility of obtaining atomic resolution structures of small molecule binding sites (e.g. Bagnéris et al., 2014; Liao et al., 2013; Yohannan et al., 2007) in the near future, more predictive ion channel drug discoveries will become possible.

Drug name	Chemical structure	Target channel	Disease target	clinical usage since
Verapamil		L-type Ca_v	Hypertension	1982
Diltiazem		L-type Ca_v	Hypertension	1982
Sotalol		hERG K_v	Arrhythmia	1992
Flecainide		$Na_v1.5$	Arrhythmia	1982
Ziconotide	Peptide	$Ca_v2.2$	Severe pain	2004
Lidocaine		Na_v	Local anaesthetic	1949
Phenytoin		Na_v	Epilepsy	1953
Tolbutamide		K_{ATP}	Diabetes	1989

Table 1- marketed ion channel drugs (modified from Bagal et al., 2013).

Despite this success for the treatment of various diseases, many challenges remain. Of particular importance, ion channels are not isolated ion-conduction pores in the membrane of every cell, but it is becoming more and more clear that these channels are part of large signalling complexes (Lee et al., 2014).

Another challenge that remains concerns the precise determination of the ion channel types, underlying certain currents in native tissues, due to the ability of many α-subunits to heteromultimerize. These heteromeric channels have widely different biophysical and pharmacological properties. Next to the molecular diversity ion channel are further modified via phosphorylation, ubiquitylation, palmitoylation etcetera. Thus, designing modulators that are able to discriminate between these often closely related channel types, remains a considerable challenge in drug discovery. Excellent reviews and books (e.g. Bagal et al., 2013; Cox and Gosling, 2014) about ion channel drug discovery have been written and are beyond the scope of this introduction.

One ion channel that has received particular attention in the drug discovery field, due to its off-target pharmacology, is the hERG K^+ channel, which is a major focus of my current research (**papers #2, #5, #6, #7 and #8** (Stary et al., 2010; Knape et al., 2011; Garg et al., 2011; Garg et al., 2013; Linder et al., 2014). Thus, a brief introduction about this channel is given below. For an excellent in-depth review see Vandenberg et al., 2012.

2.1 The hERG K^+ channel

The hERG (human ether-à-go-go related K^+ channel gene) encodes for the K^+ channel α- subunit $K_v11.1$. This channel is involved in normal repolarization of the cardiac action potential. Disruption of normal channel function due to inherited mutations or more commonly due to block by structurally diverse medications can lead to long QT syndrome and sudden cardiac death (Chiang and Roden, 2000; Keating and Sanguinetti, 2001; Sanguinetti et al., 1995). Inhibition of hERG by non-cardiac medications (i.e. antihistamines, antibiotics, antipsychotics) constitutes a serious challenge, since unintended block can lead to life-threatening arrhythmias. Consequently, intense efforts have been invested into the understanding of the molecular and structural basis of hERG channel drug interactions. Alanine scans of the inner pore helices, as well as parts of the selectivity filter region led to the identification of putative binding residues in these segments

(Mitcheson et al., 2000a). The strongest effect on drug block was observed when mutating two aromatic residues Y652 and F656 to alanine. For example, these mutants reduced the affinity for the well-known hERG blocker MK-499 by 94-fold and 650-fold, respectively (Mitcheson et al., 2000a). Comparably strong effects have been observed for many structurally diverse compounds. In combination with computational models this led to the well accepted hypothesis that many compounds bind in the inner cavity of hERG and form π- π or cation- π interactions with Y652 and F656 (Farid et al., 2006; Fernandez et al., 2004).

A major challenge, towards understanding drug block in the hERG K^+ channel in detail comes from the fact that currently no experimentally determined 3D structures of the pore domain of this channel is available. Thus modelling approaches are needed to obtain insights into the structure of these channels and to derive at mechanistic models explaining drug block. Consequently, many comparative structure modelling efforts have been published in the literature over the last years (see **paper #2**, Stary et al., 2010). Due to the low sequence identity between currently available 3D structures of K^+ channels and hERG, various models, based on different templates and alignments, have been published. In particular, no consensus of the alignment of the outer pore helix (S5) was achieved and published models showed considerable variation in this region. To achieve a consensus model of this important region (see also **paper #7**, Garg et al., 2011) we performed systematic analyses with seven homology models, based on different S5 alignments. By combining geometry based quality assessment tools, molecular dynamics simulations and experimental data we were able to identify a consensus model, which fulfils all quality criteria. This model was subsequently used to investigate drug interactions in hERG in detail (see **papers #5, #6, #7 and #8**).

In **paper # 5**, (Knape et al., 2011) we investigated possible structural changes upon mutation of the two aromatic binding residues to alanine by applying *in silico* mutations combined with molecular dynamics simulations. Our simulations reveal that π- π stacking interactions between the aromatic side chains of Y652 and F656 are important for stabilizing a cavity facing orientation of these two binding determinants. Upon mutation of Y652 to alanine subtle structural rearrangements of the F656 side chain were observed. To investigate if these geometry changes influence drug interactions we performed docking studies and subsequent MD simulations with selected blockers. Our study indicates that depending on the geometry of the blocker (flexible elongated vs. compact rigid) subtle changes in the binding site can influence the number of favourable drug-channel interactions. Specifically our investigations led to the hypothesis that flexible, elongated compounds, such as cisapride should be more sensitive to variations in binding site geometry, compared to compact, rigid molecules such as bepridil. In agreement with this hypothesis, experimental data indicates that compounds such as bepridil are less sensitive to mutations in position Y652.

Another interesting feature of hERG – drug interactions concerns the kinetics of drug block and dissociation. Of particular interest is a phenomenon called "drug trapping", which describes the phenomenon that for certain drugs dissociation is prevented by closure of the activation gate (Mitcheson et al., 2000b; Stork et al., 2007). Drug trapping has been demonstrated first in squid axons for quaternary ammonium ion compound by Armstrong in 1971(Armstrong, 1971). A possible clinical significance of this phenomenon in relation to pro-arrhythmic risk has been described recently (Di Veroli et al., 2013). In **paper #6** we describe novel insights into the trapping phenomenon by applying the essential dynamics method as described in detail in **paper #4** and the well-studied small molecule blocker tetrabutylammonium (TBA). The most interesting, novel finding of our study was that TBA affected the structure of the hERG cavity, in particular the dynamics of F656. This is in line with experimental studies, suggesting that F656 might act as physical barrier for drug dissociation of trapped compounds (Karczewski et al., 2009).

A few years ago, hERG K^+ channel agonists were discovered accidentally, during mandatory preclinical screens for hERG block. hERG channel activators shorten cardiac action potentials by altering channel gating via different mechanisms. Mechanisms of activator action include suppression of C-type inactivation, slowed deactivation, increase of channel open probability or a combination of these effects. These novel compounds have an interesting therapeutic potential and might constitute an interesting approach to treat long QT syndrome (Sanguinetti, 2014). Further, these compounds might be useful tools to investigate gating mechanisms in hERG channels in more detail. Our own contribution to this field of study is summarized in **papers #7 and #8**, where we provide detailed insights into the structural and molecular basis of hERG activator 3-nitro-N-(4-phenoxyphenyl) benzamide (ICA). ICA increases hERG outward current more than 10-fold by inducing profound positive shifts in voltage dependent C-type inactivation (Sanguinetti, 2014). By combining scanning mutagenesis and molecular modelling the molecular determinants of ICA action and a putative binding site at the interface between S6 segments is proposed. Remarkably, one mutation in S6 (A653M) enabled switching of the activator effect of ICA to an inhibitor, suggesting close overlap between the binding sites of activators and blockers. In **paper #8** we used the same approach to test the effects of this activator on the closely related EAG channels, since these channels differ with respect to their inactivation properties. While hERG K^+ channels inactivate very rapidly in a voltage dependent manner (ms time scale), inactivation in EAG K^+ channels evolves very slowly, with minimal reduction of outward currents. Interestingly, our study reveals that ICA inhibits outward K^+ current in hEAG, by enhancing slow inactivation. Even more interesting, our mutational analysis combined with molecular modelling suggests that ICA bindings to a common binding site, despite inducing opposite effects on inactivation. Further studies to investigate the structural basis of the poorly understood inactivation mechanism in K_v channels are currently in progress in our lab.

3 References

Adelman, J.P., Bond, C.T., Pessia, M., and Maylie, J. (1995). Episodic ataxia results from voltage-dependent potassium channels with altered functions. Neuron *15*, 1449–1454.

Aqvist, J., and Luzhkov, V. (2000). Ion permeation mechanism of the potassium channel. Nature *404*, 881–884.

Armstrong, C.M. (1971). Interaction of tetraethylammonium ion derivatives with the potassium channels of giant axons. J. Gen. Physiol. *58*, 413–437.

Bagal, S.K., Brown, A.D., Cox, P.J., Omoto, K., Owen, R.M., Pryde, D.C., Sidders, B., Skerratt, S.E., Stevens, E.B., Storer, R.I., et al. (2013). Ion channels as therapeutic targets: a drug discovery perspective. J. Med. Chem. *56*, 593–624.

Bagnéris, C., DeCaen, P.G., Hall, B.A., Naylor, C.E., Clapham, D.E., Kay, C.W.M., and Wallace, B.A. (2013). Role of the C-terminal domain in the structure and function of tetrameric sodium channels. Nat. Commun. *4*.

Bagnéris, C., DeCaen, P.G., Naylor, C.E., Pryde, D.C., Nobeli, I., Clapham, D.E., and Wallace, B.A. (2014). Prokaryotic NavMs channel as a structural and functional model for eukaryotic sodium channel antagonism. Proc. Natl. Acad. Sci. U. S. A. *111*, 8428–8433.

Bavro, V.N., De Zorzi, R., Schmidt, M.R., Muniz, J.R.C., Zubcevic, L., Sansom, M.S.P., Vénien-Bryan, C., and Tucker, S.J. (2012). Structure of a KirBac potassium channel with an open bundle crossing indicates a mechanism of channel gating. Nat. Struct. Mol. Biol. *19*, 158–163.

Beyl, S., Depil, K., Hohaus, A., Stary-Weinzinger, A., Linder, T., Timin, E., and Hering, S. (2012). Neutralisation of a single voltage sensor affects gating determinants in all four pore-forming S6 segments of CaV1.2: a cooperative gating model. Pflüg. Arch. - Eur. J. Physiol. *464*, 391–401.

Beyl, S., Kügler, P., Hohaus, A., Depil, K., Hering, S., and Timin, E. (2014). Methods for quantification of pore-voltage sensor interaction in Ca(V)1.2. Pflüg. Arch. Eur. J. Physiol. *466*, 265–274.

Biggin, P.C., and Sansom, M.S.P. (2002). Open-state models of a potassium channel. Biophys. J. *83*, 1867–1876.

Boczek, N.J., Miller, E.M., Ye, D., Nesterenko, V.V., Tester, D.J., Antzelevitch, C., Czosek, R.J., Ackerman, M.J., and Ware, S.M. (2014). Novel Timothy syndrome mutation leading to increase in CACNA1C window current. Heart Rhythm Off. J. Heart Rhythm Soc.

Boiteux, C., Vorobyov, I., French, R.J., French, C., Yarov-Yarovoy, V., and Allen, T.W. (2014). Local anesthetic and antiepileptic drug access and binding to a bacterial voltage-gated sodium channel. Proc. Natl. Acad. Sci. U. S. A. *111*, 13057–13062.

Boukharta, L., Keränen, H., Stary-Weinzinger, A., Wallin, G., de Groot, B.L., and Aqvist, J. (2011). Computer simulations of structure-activity relationships for HERG channel blockers. Biochemistry (Mosc.) *50*, 6146–6156.

Brohawn, S.G., del Mármol, J., and MacKinnon, R. (2012). Crystal structure of the human K2P TRAAK, a lipid- and mechano-sensitive K+ ion channel. Science *335*, 436–441.

Cannon, S.C. (2007). Physiologic principles underlying ion channelopathies. Neurotherapeutics *4*, 174–183.

Carnevale, V., Treptow, W., and Klein, M.L. (2011). Sodium Ion Binding Sites and Hydration in the Lumen of a Bacterial Ion Channel from Molecular Dynamics Simulations. J. Phys. Chem. Lett. *2*, 2504–2508.

Catterall, W.A. (1986). Voltage-dependent gating of sodium channels: correlating structure and function. Trends Neurosci. *9*, 7–10.

Catterall, W.A. (2012). Voltage-gated sodium channels at 60: structure, function and pathophysiology. J. Physiol. *590*, 2577–2589.

Chen, J., Seebohm, G., and Sanguinetti, M.C. (2002). Position of aromatic residues in the S6 domain, not inactivation, dictates cisapride sensitivity of HERG and eag potassium channels. Proc. Natl. Acad. Sci. U. S. A. *99*, 12461–12466.

Chen, J., Luan, Y., Yu, R., Zhang, Z., Zhang, J., and Wang, W. (2014). Transient receptor potential (TRP) channels, promising potential diagnostic and therapeutic tools for cancer. Biosci. Trends *8*, 1–10.

Chiang, C.E., and Roden, D.M. (2000). The long QT syndromes: genetic basis and clinical implications. J. Am. Coll. Cardiol. *36*, 1–12.

Clarke, O.B., Caputo, A.T., Hill, A.P., Vandenberg, J.I., Smith, B.J., and Gulbis, J.M. (2010). Domain reorientation and rotation of an intracellular assembly regulate conduction in Kir potassium channels. Cell *141*, 1018–1029.

Cole, K.S. (1979). Mostly membranes (Kenneth S. Cole). Annu. Rev. Physiol. *41*, 1–24.

Colombini, M. (2012). VDAC structure, selectivity, and dynamics. Biochim. Biophys. Acta *1818*, 1457–1465.

Corry, B., and Thomas, M. (2012). Mechanism of ion permeation and selectivity in a voltage gated sodium channel. J. Am. Chem. Soc. *134*, 1840–1846.

Cox, B., and Gosling, M. (2014). Ion Channel Drug Discovery (Royal Society of Chemistry).

Cuello, L.G., Jogini, V., Cortes, D.M., and Perozo, E. (2010). Structural mechanism of C-type inactivation in K(+) channels. Nature *466*, 203–208.

Debenham, J.S., Graham, T.H., Verras, A., Zhang, Y., Clements, M.J., Kuethe, J.T., Madsen-Duggan, C., Liu, W., Bhatt, U.R., Chen, D., et al. (2013). Discovery and optimization of orally active cyclohexane-based prolylcarboxypeptidase (PrCP) inhibitors. Bioorg. Med. Chem. Lett. *23*, 6228–6233.

DeCaen, P.G., Yarov-Yarovoy, V., Sharp, E.M., Scheuer, T., and Catterall, W.A. (2009). Sequential formation of ion pairs during activation of a sodium channel voltage sensor. Proc. Natl. Acad. Sci. *106*, 22498–22503.

Denning, E.J., and Woolf, T.B. (2010). Cooperative nature of gating transitions in K(+) channels as seen from dynamic importance sampling calculations. Proteins *78*, 1105–1119.

Depil, K., Beyl, S., Stary-Weinzinger, A., Hohaus, A., Timin, E., and Hering, S. (2011). Timothy mutation disrupts the link between activation and inactivation in Ca(V)1.2 protein. J. Biol. Chem. *286*, 31557–31564.

Doyle, D.A., Morais Cabral, J., Pfuetzner, R.A., Kuo, A., Gulbis, J.M., Cohen, S.L., Chait, B.T., and MacKinnon, R. (1998). The structure of the potassium channel: molecular basis of K+ conduction and selectivity. Science *280*, 69–77.

Efremov, R.G., Leitner, A., Aebersold, R., and Raunser, S. (2014). Architecture and conformational switch mechanism of the ryanodine receptor. Nature.

Farid, R., Day, T., Friesner, R.A., and Pearlstein, R.A. (2006). New insights about HERG blockade obtained from protein modelling, potential energy mapping, and docking studies. Bioorg. Med. Chem. *14*, 3160–3173.

Fernandez, D., Ghanta, A., Kauffman, G.W., and Sanguinetti, M.C. (2004). Physicochemical features of the HERG channel drug binding site. J. Biol. Chem. *279*, 10120–10127.

Fraser, S.P., Ozerlat-Gunduz, I., Brackenbury, W.J., Fitzgerald, E.M., Campbell, T.M., Coombes, R.C., and Djamgoz, M.B.A. (2014). Regulation of voltage-gated sodium channel expression in cancer: hormones, growth factors and auto-regulation. Philos. Trans. R. Soc. Lond. B. Biol. Sci. *369*, 20130105.

Furini, S., and Domene, C. (2012). On Conduction in a Bacterial Sodium Channel. PLoS Comput Biol *8*, e1002476.

Furini, S., and Domene, C. (2013). K(+) and Na(+) conduction in selective and nonselective ion channels via molecular dynamics simulations. Biophys. J. *105*, 1737–1745.

Gao, Y., Xue, X., Hu, D., Liu, W., Yuan, Y., Sun, H., Li, L., Timothy, K.W., Zhang, L., Li, C., et al. (2013). Inhibition of Late Sodium Current by Mexiletine A Novel Pharmotherapeutical Approach in Timothy Syndrome. Circ. Arrhythm. Electrophysiol. *6*, 614–622.

Garg, V., Stary-Weinzinger, A., Sachse, F., and Sanguinetti, M.C. (2011). Molecular determinants for activation of human ether-à-go-go-related gene 1 potassium channels by 3-nitro-n-(4-phenoxyphenyl) benzamide. Mol. Pharmacol. *80*, 630–637.

Garg, V., Sachse, F.B., and Sanguinetti, M.C. (2012). Tuning of EAG K(+) channel inactivation: molecular determinants of amplification by mutations and a small molecule. J. Gen. Physiol. *140*, 307–324.

Garg, V., Stary-Weinzinger, A., and Sanguinetti, M.C. (2013). ICA-105574 interacts with a common binding site to elicit opposite effects on inactivation gating of EAG and ERG potassium channels. Mol. Pharmacol. *83*, 805–813.

Guy, H.R., and Seetharamulu, P. (1986). Molecular model of the action potential sodium channel. Proc. Natl. Acad. Sci. *83*, 508–512.

Henrion, U., Renhorn, J., Börjesson, S.I., Nelson, E.M., Schwaiger, C.S., Bjelkmar, P., Wallner, B., Lindahl, E., and Elinder, F. (2012). Tracking a complete voltage-sensor cycle with metal-ion bridges. Proc. Natl. Acad. Sci. *109*, 8552–8557.

Hille, B. (2001). Ionic Channels of Excitable Membranes (Sunderland, Mass: Palgrave Macmillan).

Hodgkin, A.L., and Huxley, A.F. (1952a). A quantitative description of membrane current and its application to conduction and excitation in nerve. J. Physiol. *117*, 500–544.

Hodgkin, A.L., and Huxley, A.F. (1952b). Propagation of electrical signals along giant nerve fibers. Proc. R. Soc. Lond. Ser. B Contain. Pap. Biol. Character R. Soc. G. B. *140*, 177–183.

Hoomann, T., Jahnke, N., Horner, A., Keller, S., and Pohl, P. (2013). Filter gate closure inhibits ion but not water transport through potassium channels. Proc. Natl. Acad. Sci. *110*, 10842–10847.

Hoshi, T., and Armstrong, C.M. (2013). C-type inactivation of voltage-gated K+ channels: pore constriction or dilation? J. Gen. Physiol. *141*, 151–160.

Huang, X., and Jan, L.Y. (2014). Targeting potassium channels in cancer. J. Cell Biol. *206*, 151–162.

Irie, K., Kitagawa, K., Nagura, H., Imai, T., Shimomura, T., and Fujiyoshi, Y. (2010). Comparative study of the gating motif and C-type inactivation in prokaryotic voltage-gated sodium channels. J. Biol. Chem. *285*, 3685–3694.

Jensen, MO., Jogini, V., Borhani, DW., Leffler AE., Dror, RO., Shaw, DE. (2012). Mechanism of voltage gating in potassium channels. Science 336, 6078, 229-233.

Jiang, Y., Lee, A., Chen, J., Ruta, V., Cadene, M., Chait, B.T., and MacKinnon, R. (2003). X-ray structure of a voltage-dependent K+ channel. Nature *423*, 33–41.

Karczewski, J., Wang, J., Kane, S.A., Kiss, L., Koblan, K.S., Culberson, J.C., and Spencer, R.H. (2009). Analogs of MK-499 are differentially affected by a mutation in the S6 domain of the hERG K+ channel. Biochem. Pharmacol. *77*, 1602–1611.

Ke, S., Zangerl, E.-M., and Stary-Weinzinger, A. (2013). Distinct interactions of Na+ and Ca2+ ions with the selectivity filter of the bacterial sodium channel Na(V)Ab. Biochem. Biophys. Res. Commun. *430*, 1272–1276.

Ke, S., Timin, E.N., and Stary-Weinzinger, A. (2014). Different inward and outward conduction mechanisms in NaVMs suggested by molecular dynamics simulations. PLoS Comput. Biol. *10*, e1003746.

Keating, M.T., and Sanguinetti, M.C. (2001). Molecular and cellular mechanisms of cardiac arrhythmias. Cell *104*, 569–580.

Kelly, B.L., and Gross, A. (2003). Potassium channel gating observed with site-directed mass tagging. Nat. Struct. Biol. *10*, 280–284.

Kim, J.-B. (2014). Channelopathies. Korean J. Pediatr. *57*, 1–18.

Knape, K., Linder, T., Wolschann, P., Beyer, A., and Stary-Weinzinger, A. (2011). In silico analysis of conformational changes induced by mutation of aromatic binding residues: consequences for drug binding in the hERG K+ channel. PloS One 6, e28778.

Köpfer, D.A., Song, C., Gruene, T., Sheldrick, G.M., Zachariae, U., and de Groot, B.L. (2014). Ion permeation in K^+ channels occurs by direct Coulomb knock-on. Science 346, 352–355.

Kurata, H.T., and Fedida, D. (2006). A structural interpretation of voltage-gated potassium channel inactivation. Prog. Biophys. Mol. Biol. 92, 185–208.

Kutzner, C., Grubmüller, H., de Groot, B.L., and Zachariae, U. (2011). Computational electrophysiology: the molecular dynamics of ion channel permeation and selectivity in atomistic detail. Biophys. J. 101, 809–817.

Lacroix, J.J., Pless, S.A., Maragliano, L., Campos, F.V., Galpin, J.D., Ahern, C.A., Roux, B., and Bezanilla, F. (2012). Intermediate state trapping of a voltage sensor. J. Gen. Physiol. 140, 635–652.

Lee, A., Fakler, B., Kaczmarek, L.K., and Isom, L.L. (2014). More than a pore: ion channel signaling complexes. J. Neurosci. Off. J. Soc. Neurosci. 34, 15159–15169.

Lemoine, D., Jiang, R., Taly, A., Chataigneau, T., Specht, A., and Grutter, T. (2012). Ligand-gated ion channels: new insights into neurological disorders and ligand recognition. Chem. Rev. 112, 6285–6318.

Liao, M., Cao, E., Julius, D., and Cheng, Y. (2013). Structure of the TRPV1 ion channel determined by electron cryo-microscopy. Nature 504, 107–112.

Linder, T., de Groot, B.L., and Stary-Weinzinger, A. (2013). Probing the energy landscape of activation gating of the bacterial potassium channel KcsA. PLoS Comput. Biol. 9, e1003058.

Linder, T., Saxena, P., Timin, E., Hering, S., and Stary-Weinzinger, A. (2014). Structural Insights into Trapping and Dissociation of Small Molecules in K(+) Channels. J. Chem. Inf. Model.

Long, S.B., Campbell, E.B., and Mackinnon, R. (2005). Crystal structure of a mammalian voltage-dependent Shaker family K+ channel. Science 309, 897–903.

Martin, L.J., and Corry, B. (2014). Locating the route of entry and binding sites of benzocaine and phenytoin in a bacterial voltage gated sodium channel. PLoS Comput. Biol. 10, e1003688.

Masetti, M., Cavalli, A., and Recanatini, M. (2008). Modelling the hERG potassium channel in a phospholipid bilayer: Molecular dynamics and drug docking studies. J. Comput. Chem. 29, 795–808.

McCusker, E.C., Bagnéris, C., Naylor, C.E., Cole, A.R., D'Avanzo, N., Nichols, C.G., and Wallace, B.A. (2012). Structure of a bacterial voltage-gated sodium channel pore reveals mechanisms of opening and closing. Nat. Commun. 3, 1102.

Miller, A.N., and Long, S.B. (2012). Crystal structure of the human two-pore domain potassium channel K2P1. Science 335, 432–436.

Mitcheson, J.S., Chen, J., Lin, M., Culberson, C., and Sanguinetti, M.C. (2000a). A structural basis for drug-induced long QT syndrome. Proc. Natl. Acad. Sci. U. S. A. 97, 12329–12333.

Mitcheson, J.S., Chen, J., and Sanguinetti, M.C. (2000b). Trapping of a methanesulfonanilide by closure of the HERG potassium channel activation gate. J. Gen. Physiol. 115, 229–240.

Morais-Cabral, J.H., Zhou, Y., and MacKinnon, R. (2001). Energetic optimization of ion conduction rate by the K+ selectivity filter. Nature *414*, 37–42.

Moreau, A., Gosselin-Badaroudine, P., and Chahine, M. (2014). Biophysics, pathophysiology, and pharmacology of ion channel gating pores. Front. Pharmacol. *5*, 53.

Neher, E., and Sakmann, B. (1976). Single-channel currents recorded from membrane of denervated frog muscle fibres. Nature *260*, 799–802.

Noda, M., Takahashi, H., Tanabe, T., Toyosato, M., Furutani, Y., Hirose, T., Asai, M., Inayama, S., Miyata, T., and Numa, S. (1982). Primary structure of alpha-subunit precursor of Torpedo californica acetylcholine receptor deduced from cDNA sequence. Nature *299*, 793–797.

Olson, T.M., and Terzic, A. (2010). Human K(ATP) channelopathies: diseases of metabolic homeostasis. Pflüg. Arch. Eur. J. Physiol. *460*, 295–306.

Österberg, F., and Åqvist, J. (2005). Exploring blocker binding to a homology model of the open hERG K+ channel using docking and molecular dynamics methods. FEBS Lett. *579*, 2939–2944.

Ostmeyer, J., Chakrapani, S., Pan, A.C., Perozo, E., and Roux, B. (2013). Recovery from slow inactivation in K+ channels is controlled by water molecules. Nature *501*, 121–124.

Overington, J.P., Al-Lazikani, B., and Hopkins, A.L. (2006). How many drug targets are there? Nat. Rev. Drug Discov. *5*, 993–996.

Pan, A.C., Cuello, L.G., Perozo, E., and Roux, B. (2011). Thermodynamic coupling between activation and inactivation gating in potassium channels revealed by free energy molecular dynamics simulations. J. Gen. Physiol. *138*, 571–580.

Pardo, L.A., and Stühmer, W. (2014). The roles of K(+) channels in cancer. Nat. Rev. Cancer *14*, 39–48.

Pawson, A.J., Sharman, J.L., Benson, H.E., Faccenda, E., Alexander, S.P.H., Buneman, O.P., Davenport, A.P., McGrath, J.C., Peters, J.A., Southan, C., et al. (2014). The IUPHAR/BPS Guide to PHARMACOLOGY: an expert-driven knowledgebase of drug targets and their ligands. Nucleic Acids Res. *42*, D1098–D1106.

Payandeh, J., Scheuer, T., Zheng, N., and Catterall, W.A. (2011). The crystal structure of a voltage-gated sodium channel. Nature *475*, 353–358.

Payandeh, J., Gamal El-Din, T.M., Scheuer, T., Zheng, N., and Catterall, W.A. (2012). Crystal structure of a voltage-gated sodium channel in two potentially inactivated states. Nature *486*, 135–139.

Perozo, E., Cortes, D.M., and Cuello, L.G. (1999). Structural rearrangements underlying K+-channel activation gating. Science *285*, 73–78.

Peters, C.J., Fedida, D., and Accili, E.A. (2013). Allosteric coupling of the inner activation gate to the outer pore of a potassium channel. Sci. Rep. *3*, 3025.

Poulsen, H., and Nissen, P. (2012). The Inner Workings of a Dynamic Duo. Science *335*, 416–417.

Raju, S.G., Barber, A.F., LeBard, D.N., Klein, M.L., and Carnevale, V. (2013). Exploring volatile general anesthetic binding to a closed membrane-bound bacterial voltage-gated sodium channel via computation. PLoS Comput. Biol. *9*, e1003090.

Rapedius, M., Schmidt, M.R., Sharma, C., Stansfeld, P.J., Sansom, M.S.P., Baukrowitz, T., and Tucker, S.J. (2012). State-independent intracellular access of quaternary ammonium blockers to the pore of TREK-1. Channels Austin Tex 6, 473–478.

Sánchez-Chapula, J.A., Ferrer, T., Navarro-Polanco, R.A., and Sanguinetti, M.C. (2003). Voltage-dependent profile of human ether-a-go-go-related gene channel block is influenced by a single residue in the S6 transmembrane domain. Mol. Pharmacol. 63, 1051–1058.

Sanguinetti, M.C. (2014). HERG1 channel agonists and cardiac arrhythmia. Curr. Opin. Pharmacol. 15, 22–27.

Sanguinetti, M.C., Jiang, C., Curran, M.E., and Keating, M.T. (1995). A mechanistic link between an inherited and an acquired cardiac arrhythmia: HERG encodes the IKr potassium channel. Cell 81, 299–307.

Shaya, D., Findeisen, F., Abderemane-Ali, F., Arrigoni, C., Wong, S., Nurva, S.R., Loussouarn, G., and Minor, D.L. (2014). Structure of a prokaryotic sodium channel pore reveals essential gating elements and an outer ion binding site common to eukaryotic channels. J. Mol. Biol. 426, 467–483.

Shimizu, H., Iwamoto, M., Konno, T., Nihei, A., Sasaki, Y.C., and Oiki, S. (2008). Global twisting motion of single molecular KcsA potassium channel upon gating. Cell 132, 67–78.

Shrivastava, I.H., and Bahar, I. (2006). Common mechanism of pore opening shared by five different potassium channels. Biophys. J. 90, 3929–3940.

Splawski, I., Timothy, K.W., Priori, S.G., Napolitano, C., and Bloise, R. (1993). Timothy Syndrome. In GeneReviews(®), R.A. Pagon, M.P. Adam, H.H. Ardinger, T.D. Bird, C.R. Dolan, C.-T. Fong, R.J. Smith, and K. Stephens, eds. (Seattle (WA): University of Washington, Seattle),.

Splawski, I., Timothy, K.W., Sharpe, L.M., Decher, N., Kumar, P., Bloise, R., Napolitano, C., Schwartz, P.J., Joseph, R.M., Condouris, K., et al. (2004). CaV1.2 Calcium Channel Dysfunction Causes a Multisystem Disorder Including Arrhythmia and Autism. Cell 119, 19–31.

Stansfeld, P.J., Gedeck, P., Gosling, M., Cox, B., Mitcheson, J.S., and Sutcliffe, M.J. (2007). Drug block of the hERG potassium channel: Insight from modelling. Proteins Struct. Funct. Bioinforma. 68, 568–580.

Stary, A., Wacker, S.J., Boukharta, L., Zachariae, U., Karimi-Nejad, Y., Aqvist, J., Vriend, G., and de Groot, B.L. (2010). Toward a consensus model of the HERG potassium channel. ChemMedChem 5, 455–467.

Stock, L., Delemotte, L., Carnevale, V., Treptow, W., and Klein, M.L. (2013). Conduction in a Biological Sodium Selective Channel. J. Phys. Chem. B 117, 3782–3789.

Stölting, G., Fischer, M., and Fahlke, C. (2014). CLC channel function and dysfunction in health and disease. Front. Physiol. 5, 378.

Stork, D., Timin, E.N., Berjukow, S., Huber, C., Hohaus, A., Auer, M., and Hering, S. (2007). State dependent dissociation of HERG channel inhibitors. Br. J. Pharmacol. 151, 1368–1376.

Tao, X., Avalos, J.L., Chen, J., and MacKinnon, R. (2009). Crystal structure of the eukaryotic strong inward-rectifier K+ channel Kir2.2 at 3.1 A resolution. Science 326, 1668–1674.

Tikhonov, D.B., and Zhorov, B.S. (2004). In silico activation of KcsA K+ channel by lateral forces applied to the C-termini of inner helices. Biophys. J. 87, 1526–1536.

Tsai, C.-J., Tani, K., Irie, K., Hiroaki, Y., Shimomura, T., McMillan, D.G., Cook, G.M., Schertler, G.F.X., Fujiyoshi, Y., and Li, X.-D. (2013). Two alternative conformations of a voltage-gated sodium channel. J. Mol. Biol. *425*, 4074–4088.

Tseng, G.-N., Sonawane, K.D., Korolkova, Y.V., Zhang, M., Liu, J., Grishin, E.V., and Guy, H.R. (2007). Probing the Outer Mouth Structure of the hERG Channel with Peptide Toxin Footprinting and Molecular Modelling. Biophys. J. *92*, 3524–3540.

Uysal, S., Vásquez, V., Tereshko, V., Esaki, K., Fellouse, F.A., Sidhu, S.S., Koide, S., Perozo, E., and Kossiakoff, A. (2009). Crystal structure of full-length KcsA in its closed conformation. Proc. Natl. Acad. Sci. U. S. A. *106*, 6644–6649.

Vandenberg, J.I., Perry, M.D., Perrin, M.J., Mann, S.A., Ke, Y., and Hill, A.P. (2012). hERG K(+) channels: structure, function, and clinical significance. Physiol. Rev. *92*, 1393–1478.

Vargas, E., Yarov-Yarovoy, V., Khalili-Araghi, F., Catterall, W.A., Klein, M.L., Tarek, M., Lindahl, E., Schulten, K., Perozo, E., Bezanilla, F., et al. (2012). An emerging consensus on voltage-dependent gating from computational modelling and molecular dynamics simulations. J. Gen. Physiol. *140*, 587–594.

DI Veroli, G.Y., Davies, M.R., Zhang, H., Abi-Gerges, N., and Boyett, M.R. (2013). hERG Inhibitors With Similar Potency But Different Binding Kinetics Do Not Pose the Same Proarrhythmic Risk: Implications for Drug Safety Assessment. J. Cardiovasc. Electrophysiol.

Wang, Y., Wilson, S.M., Brittain, J.M., Ripsch, M.S., Salomé, C., Park, K.D., White, F.A., Khanna, R., and Kohn, H. (2011). Merging Structural Motifs of Functionalized Amino Acids and α-Aminoamides Results in Novel Anticonvulsant Compounds with Significant Effects on Slow and Fast Inactivation of Voltage-gated Sodium Channels and in the Treatment of Neuropathic Pain. ACS Chem. Neurosci. *2*, 317–322.

Xia, M., Liu, H., Li, Y., Yan, N., and Gong, H. (2013). The Mechanism of Na+/K+ Selectivity in Mammalian Voltage-Gated Sodium Channels Based on Molecular Dynamics Simulation. Biophys. J. *104*, 2401–2409.

Yohannan, S., Hu, Y., and Zhou, Y. (2007). Crystallographic study of the tetrabutylammonium block to the KcsA K+ channel. J. Mol. Biol. *366*, 806–814.

Zachariae, U., Giordanetto, F., and Leach, A.G. (2009). Side chain flexibilities in the human ether-a-go-go related gene potassium channel (hERG) together with matched-pair binding studies suggest a new binding mode for channel blockers. J. Med. Chem. *52*, 4266–4276.

Zhang, X., Ren, W., DeCaen, P., Yan, C., Tao, X., Tang, L., Wang, J., Hasegawa, K., Kumasaka, T., He, J., et al. (2012). Crystal structure of an orthologue of the NaChBac voltage-gated sodium channel. Nature *486*, 130–134.

Zhang, X., Xia, M., Li, Y., Liu, H., Jiang, X., Ren, W., Wu, J., DeCaen, P., Yu, F., Huang, S., et al. (2013). Analysis of the selectivity filter of the voltage-gated sodium channel NavRh. Cell Res. *23*, 409–422.

Zhou, Y., Morais-Cabral, J.H., Kaufman, A., and MacKinnon, R. (2001). Chemistry of ion coordination and hydration revealed by a K+ channel-Fab complex at 2.0 A resolution. Nature *414*, 43–48.

Zimmer, J., Doyle, D.A., and Grossmann, J.G. (2006). Structural characterization and pH-induced conformational transition of full-length KcsA. Biophys. J. *90*, 1752–1766.

4 Short summaries of papers described in this habilitation thesis

On the following pages short summaries about the eight papers described in this habilitation thesis are given. The actual paper reprints are attached below.

4.1 Timothy Mutation Disrupts the Link between Activation and Inactivation in Cav1.2 Protein

Katrin Depil[#], Stanislav Beyl[#], Anna Stary-Weinzinger[#], Annette Hohaus, Eugen Timin, Steffen Hering

Journal of Biological Chemistry, **2011**, 286,36, 31557-31564

Summary:
Structure modelling for voltage gated calcium channel members is even more challenging than developing models for K^+ channels, such as hERG, since experimental structure information is missing for the whole family. At the time of publication of this manuscript only a handful of K^+ channels structures, were available. Unfortunately these channels are very distantly related, resulting in sequence identities way below 30%, rendering modelling challenging. Nevertheless, the overall architecture might be similar enough, to obtain low resolution models. In this paper we resorted to identification of conserved structure motifs to aid in model construction. Applying bioinformatics methods, we identified a highly conserved structure motif (G/A/G/A) in homologous positions of all four pore forming S6 segments. Interestingly, a severe calcium channelopathy, the so-called Timothy-syndrome (TS), which is a multiorgan dysfunction with physical, neurological and developmental defects, is located exactly in this structure motif. Using a combination of site-directed mutagenesis, electrophysiology measurements and structure modelling, we investigated the importance of this motif for channel gating and structural stability of the closed channel gate. Our analyses suggest that the TS glycine is part of a tightly packed "small amino acid structure motif" essential for normal channel gating.

Further studies based on this paper:
- The importance on gating of this motif is supported by recent work from our group (Beyl et al., 2012, 2014), suggesting that the G/A/G/A positions are unique with respect to interacting (possibly via allosteric mechanisms) with the voltage sensing domain(s) of the channel
- The importance of the G/A/G/A motif for gating was confirmed by our gating simulations study in KcsA (**paper #4**; Linder et al., 2013)
- Due to the fact that in 2012 several crystal structures of the closer related bacterial Na_v channels (20 - 30 % SI) became available, new generation models should be used for structural analyses in the future.

4.2 Toward a Consensus Model of the hERG Potassium channel

Anna Stary, Sören J. Wacker, Lars Boukharta, Ulrich Zachariae, Yasmin Karimi-Nejad, Johan Åqvist, Gert Vriend, Bert L. de Groot.

ChemMedChem **2010**, 5, 455-467

Summary:
In the absence of experimental structure data, understanding hERG potassium channel malfunction, especially unwanted inhibition by certain drugs, relies on homology models based on homologous K^+ channels, for which x-ray structures have been solved. Owing to the low degree of sequence identity in some portions of the protein, generating accurate and reliable structure models is challenging. Given the lack of conservation in the S5 segment, homology models based on several different alignments have been described in the literature (Farid et al., 2006; Masetti et al., 2008; Österberg and Åqvist, 2005; Stansfeld et al., 2007; Tseng et al., 2007). To achieve a hERG consensus pore model, we have carried out extensive analyses on seven different homology models, combining conventional geometry/packing/normality analyses, molecular dynamics simulations and drug docking. Our study shows that it is critically important to use a combination of methods to assess the quality of homology models. Neither geometry based methods nor MD simulations alone are able to automatically identify the correct fold. Importantly poor stability in MD simulations (high RMSD values) can identify poor model quality but stable models do not automatically denote correct fold.

The herein developed consensus model was subsequently successfully used to:

- Accurately rank the binding affinities of a series of sertindole analogues (Boukharta et al., 2011)
- The alignment of S5 was indirectly validated by our subsequent studies on hERG activator ICA, because only in our consensus model residue F557, which is the most important binding determinant, is oriented towards the activator binding site (**paper #7**, Garg et al., 2011)
- Provide a structural interpretation for the modulatory effect of Y652A mutant (**paper #5**, Knape et al., 2011)

- Used by Merck Research Laboratories (Debenham et al., 2013) to investigate hERG interactions of cyclohexane-based prolylcarboxypeptidase inhibitors

4.3 Different inward and outward conduction mechanisms in NavMs suggested by molecular dynamics simulations

Song Ke, Eugen Timin, Anna Stary-Weinzinger*

PLOS Computational Biology, **2014**, 10, 7, e1003746

Summary:
Besides structural changes at the helix bundle gate, it is important to investigate the dynamics at the selectivity filter gate, implicated in inactivation gating. Changes at this "upper gate" can induce conformational changes leading to non-conductive states, termed C-type inactivation". While detailed investigations exist for ion flux, selectivity and partially inactivation on K^+ channels, the mechanisms of conductance, selectivity and filter dynamics including inactivation gating are still poorly understood in Na_v channels. With the successfully crystallization of several bacterial Na_v channels, insights into these important questions are slowly emerging. By applying double bilayer simulations with an ion gradient, we investigated the directionality of ion flux, filter dynamics and flux rates in detail. Our simulations suggest remarkable differences between ion in- and efflux. Our simulations reveal that the outward Na^+ flux is approximately 2-fold slower, compared to influx. Interestingly, filter dynamics, in particular conformational changes at the EEEE locus determine whether Na^+ efflux is possible or prevented, depending on the number of E side chains that change their conformation. It is tempting to speculate that these conformational changes initiate early steps towards C-type inactivation.

Outlook:

- The simulation protocols used in this paper provide a starting point for more in-depth investigation of slow (C-type) inactivation mechanisms in bacterial Na_v channels
- The method used in this paper is expected to be useful to investigate the influence of ion conductance on drug block

4.4 Probing the Energy Landscape of Activation Gating of the Bacterial Potassium Channel KcsA

Tobias Linder, Bert L. de Groot, Anna Stary-Weinzinger

PLOS Computational Biology, **2013**, 9, 5, e1003058

Summary:

The bacterial K^+ channel KcsA is structurally the best characterized K^+ channel available to date. The MacKinnon and Perozo labs successfully captured this channel in several conformations (Cuello et al., 2010; Doyle et al., 1998; Uysal et al., 2009), rendering it an optimal starting point to address gating conformational changes. To investigate K^+ channel gating transitions, which normally occur on the ms time scale, and are thus not accessible to conventional MD simulations, we applied the enhanced sampling technique essential dynamics (ED) simulations to facilitate transitions from closed to open channel states and vice versa. The ED technique applies knowledge from a principal components analysis to enhance sampling along a collective mode. Having atomic resolutions structures of open and closed KcsA states, allowed the calculation of the eigenvector between these two states, which enables to enforce the transition between the two states based on a single linear, 1-dimensional difference vector, but at the same time explicitly allowing all other degrees of freedom to relax continuously. Our simulations provided detailed atomistic insights into activation gating of KcsA. A two-phasic gating process, starting with local structural rearrangements followed by large structural rearrangements, leading to full channel opening, were observed in 10 independent runs. Our simulations revealed that the G/A/G/A motif and a bulky hydrophobic residue interacting with this motif identified in paper #1 indeed plays a crucial role during activation gating. In KcsA these residues correspond to A108 and F114, which are essential for the transition from a fully closed to an intermediate gating state. Importantly, the method is able to sample an intermediate x-ray structure, not included in the simulation protocol, further validating this method to correctly sample the transition pathways in ion channels.

Further studies:

- The ED method, described in this paper, was further used to investigate gating and TBA trapping in hERG (**paper #6**, Linder et al., 2014)

4.5 In silico analysis of conformational changes induced by mutation of aromatic binding residues: consequences for drug binding in the hERG K^+ channel

Kirsten Knape, Tobias Linder, Peter Wolschann, Anton Beyer, Anna Stary-Weinzinger*

PLOS ONE, **2011**, 6, 12, e28778

Summary:

Elucidating how various small molecules block the hERG K^+ channel is a major challenge for the pharmaceutical industry. It is critical to avoid unintended hERG block, since this can lead to an unacceptable high risk of drug-induced arrhythmias, and sudden death (Sanguinetti et al., 1995). Understanding the mechanisms underlying drug induced long QT syndrome, is still a major unresolved research question. In this paper we attempted to first elucidate the structural consequences of routinely performed mutagenesis studies, especially on Y652 and F656 residues, which line the inner cavity and are thought to directly interact with most pore blockers. Molecular dynamics simulations of *in silico* mutants Y652A and F656A suggest subtle changes in the binding environment due to mutation of the aromatic side chain in position 656. By subsequently applying docking and MD simulations of several well studied high affinity hERG blockers, we provide structural insights into how these local changes, induced by the mutant, might affect drug binding in hERG. Specifically, our simulations suggest that the sensitivity to mutations at position Y652 is dependent on drug geometry. This is in agreement with experimental data, revealing that rigid, bulky drug molecules such as thioridazine or bepridil are less sensitive to mutations of Y652 compared to more flexible, elongated drug molecules such as cisapride or terfenadine.

Outlook:

- These investigations might provide a starting point to address another interesting phenomenon concerning mutations of Y652 to F/A, which changes the voltage-dependence of selected drugs (Sănchez-Chapula et al., 2003)
- This study might also provide a starting point for investigating C-type inactivation in hERG, as suggested by Chen et al. (Chen et al., 2002)

4.6 Structural insights into trapping and dissociation of small molecules in K$^+$ channels

Tobias Linder, Priyanka Saxena, Eugen Timin, Steffen Hering, Anna Stary-Weinzinger*

Journal of Chemical Information and Modelling, **2014**, 54(11):3218-28.

Summary:

It is well established that drug affinities vary depending on the channel conformation. In hERG K$^+$ channels, most pore blockers can only reach their binding site in the cavity upon opening of the activation gate. Further, drug binding can be stabilized by inactivation gating, however the exact mechanisms are not understood. Another, intriguing feature, highlighting the importance of channel dynamics for drug binding is the so called "trapping phenomenon", which is characterized by capture of certain drugs behind closed channel gates. In this research project we aimed at providing structural insights into this mechanism, by combining molecular dynamics simulations and two-electrode voltage clamp studies. Our simulations provide a novel structural hypothesis for drug trapping. Specifically, MD simulations suggest that TBA (tetrabutylammonium), when bound to the cavity, changes the closing behaviour of the channel. While in the open apo hERG structure, the F656 side chain can adopt cavity facing or lining conformations (50:50 distribution), closure necessities that at least two of the four F656 side chains changes their rotameric state to a cavity lining state to enable full channel closure. If TBA is present during channel closure, the conformational change of the F656 side chain is slowed down, likely due to stabilizing interactions with the drug. Of course further experiments are needed to validate this structural hypothesis.

Outlook:

- The herein described method will be useful to further investigate the drug trapping phenomenon using more complex (non-symmetric) hERG blockers such as MK-499
- Inspires experiments such as for example testing TBA washout experiments with a F656A mutant, to see if indeed F656 constitutes a "gate" for drug trapping

4.7 Molecular determinants for activation of human ether-à-go-go-related Gene 1 Potassium channels by 3-nitro-N-(4-phenoxyphenyl) Benzamide

Vivek Garg, Anna Stary-Weinzinger, Frank Sachse, Michael C. Sanguinetti

Molecular Pharmacology, **2011**, 80, 4, 630-637

Summary:
Over the recent years, several compounds have been discovered during mandatory off-target screening for hERG channel block, which do not block hERG currents, but enhance hERG currents via various effects ranging from suppression of P-type inactivation to slowed deactivation. The therapeutic potential of such compounds lies in the fact that they could be used to shorten action potentials, which could be useful for the treatment of LQT syndrome (Sanguinetti, 2014). In this study we aimed at identifying the molecular determinants underlying the activatory effect of 3-nitro-n-(4-phenoxyphenyl)benzamide (ICA). By combining scanning mutagenesis and molecular docking, the binding determinants and putative binding site of ICA was identified. Interestingly, the identified putative binding site at the interface between two adjacent S6 segments, below the SF shares some overlap with that of known high affinity hERG blockers. In line with this finding we identified a mutant A653M in helix S6, which could switch the activity of our compound from activator to inhibitor, further supporting a close spatial relationship between activation and inhibition sites.

Outlook:
- Paves the way for further studies into the mechanism of action of this new class of modulators, which might be useful for the treatment of LQT
- Might be useful to investigate the structural basis of inactivation in hERG and hERG related K^+ channels.

4.8 ICA-105574 interacts with a common binding site to elecit opposite effects on inactivation gating of EAG and ERG Potassium channels

Vivek Garg, Anna Stary-Weinzinger, Michael C. Sanguinetti

Molecular Pharmacology, **2013**, 83, 805-813

Summary:

The aim of this work was to obtain insights into the mechanism of ICA related changes of inactivation. To address this very interesting but complex question we resorted to EAG channels, which are closely related to ERG channels, but in contrast to ERG channels show very minimal and slow inactivation. It was previously shown that ICA has opposite effects on inactivation on these two channel types (Garg et al., 2012). Thus to investigate this mechanism in detail, we again combined site-directed mutagenesis with molecular modelling investigations. Surprisingly, our study reveals common binding sites in EAG and ERG channels, despite opposite effects on inactivation. Since only three of the identified putative binding residues are different between these two channels, we attempted to replicate the ICA binding site in ERG by introducing the corresponding triple mutations. However, these mutations did not alter the functional response to ICA. These findings let us conclude that the channel-specific different modulatory effects of compound ICA, are solely due to different inactivation gating mechanisms.

Outlook:
- Studies using closely related channels might be very useful to obtain further insights into the mechanism of gating induced changes during inactivation and possible drug induced gating changes

5 Paper reprints

Paper #1

"This research was originally published in Journal of Biological Chemistry.

Katrin Depil, Stanislav Beyl, Anna Stary-Weinzinger, Annette Hohaus, Eugen Timin, Steffen Hering. Timothy Mutation Disrupts the Link between Activation and Inactivation in Cav1.2 Protein. Journal of Biological Chemistry. 2011; 286(36):31557-31564.

© the American Society for Biochemistry and Molecular Biology."

Paper #2

Copyright Wiley-VCH Verlag GmbH & Co. KGaA.
Reproduced with permission.

Papers #7, #8

Reprinted with permission of the American Society for Pharmacology and Experimental Therapeutics. All rights reserved.

5.1 Paper #1

Timothy Mutation Disrupts the Link between Activation and Inactivation in $Ca_V1.2$ Protein[*][S]

Received for publication, April 27, 2011, and in revised form, June 16, 2011 Published, JBC Papers in Press, June 17, 2011, DOI 10.1074/jbc.M111.255273

Katrin Depil[1], **Stanislav Beyl**[1], **Anna Stary-Weinzinger**[1], **Annette Hohaus, Eugen Timin, and Steffen Hering**[2]

From the Department of Pharmacology and Toxicology, University of Vienna, Althanstrasse 14, A-1090 Vienna, Austria

The Timothy syndrome mutations G402S and G406R abolish inactivation of $Ca_V1.2$ and cause multiorgan dysfunction and lethal arrhythmias. To gain insights into the consequences of the G402S mutation on structure and function of the channel, we systematically mutated the corresponding Gly-432 of the rabbit channel and applied homology modeling. All mutations of Gly-432 (G432A/M/N/V/W) diminished channel inactivation. Homology modeling revealed that Gly-432 forms part of a highly conserved structure motif (G/A/G/A) of small residues in homologous positions of all four domains (Gly-432 (IS6), Ala-780 (IIS6), Gly-1193 (IIIS6), Ala-1503 (IVS6)). Corresponding mutations in domains II, III, and IV induced, in contrast, parallel shifts of activation and inactivation curves indicating a preserved coupling between both processes. Disruption between coupling of activation and inactivation was specific for mutations of Gly-432 in domain I. Mutations of Gly-432 removed inactivation irrespective of the changes in activation. In all four domains residues G/A/G/A are in close contact with larger bulky amino acids from neighboring S6 helices. These interactions apparently provide adhesion points, thereby tightly sealing the activation gate of $Ca_V1.2$ in the closed state. Such a structural hypothesis is supported by changes in activation gating induced by mutations of the G/A/G/A residues. The structural implications for $Ca_V1.2$ activation and inactivation gating are discussed.

Timothy syndrome (TS),[3] an autosomal dominant disorder, arises from two *de novo* missense mutations, G402S and G406R, in $Ca_V1.2$ calcium channels. This channelopathy manifests itself clinically in physical, neurological, and developmental defects and autism spectrum disorders. G406R additionally causes syndactyly (1). Functional studies have shown that TS mutations G402S and G406R dramatically reduce voltage-dependent channel inactivation, resulting in sustained membrane depolarization and increased calcium entry during an action potential (1). The resulting prolongation of the QT interval can induce severe arrhythmias causing sudden death in early childhood (1). Kinetic modeling studies suggested that the increased action potential duration and enhanced calcium entry may be associated with delayed after-depolarizations in cardiac myocytes (2).

A role of G406R in calcium-dependent inactivation was established by Raybaud *et al.* (3) and Barrett and Tsien (4). Systematic mutations of Gly-436 in rabbit $Ca_V1.2$ (corresponding to Gly-406 in the human $Ca_V1.2$) to Ala, Pro, Tyr, Glu, Arg, His, Lys, or Asp all substantially slow channel inactivation (3). A large number of amino acids involved in Ca^{2+} channel inactivation have been identified, and several molecular mechanisms for this process have been proposed (5-7).

We have previously shown that mutations in the lower third of S6 segments in $Ca_V1.2$ in most cases shift the channel inactivation and activation in a coupled manner. This is evident from similar shifts of channel activation and inactivation curves (8, 9), suggesting a serial gating scheme (R)est→(O)pen→(I)nactivated where a shift in R → O transitions would automatically shift the inactivation curve. However, Raybaud *et al.* (3) reported that mutations of the TS Gly-436 that suppressed inactivation in $Ca_V1.2$ induced only minor changes in activation gating.

The functional impact and structural basis of amino acid substitutions of the TS Gly-432 (corresponding to Timothy Gly-402 in human $Ca_V1.2$) are less understood. Our homology model suggests that Gly-432 forms part of a conserved group of small amino acids, Gly-432(IS6), Ala-780(IIS6), Gly-1193(IIIS6), and Ala-1503(IVS6), near the inner channel mouth of $Ca_V1.2$ (Fig. 1), which we call the G/A/G/A motif. A corresponding Gly-657 in hERG (Human Ether-a-go-go related Gene) potassium channels was found to be important for close helix packing (10), and in KcsA, replacement of the corresponding small and hydrophobic Ala-108 by a polar serine or threonine (A108S/T) dramatically increased channel open probability (11, 12). Small gating-sensitive residues at this position thus seem to be essential for helix-helix interactions in a number of ion channels.

To clarify the role of Gly-432 as part of the apparent G/A/G/A motif of $Ca_V1.2$, we have asked several questions. First, do mutations of the homologous small residues in the other domains, Ala-780(IIS6), Gly-1193(IIIS6), and Ala-1503(IVS6), have similar effects on channel gating to substitutions of Gly-432? Second, how do amino acid substitutions in these positions affect the link between activation and inactivation? Third, is an essential role of the G/A/G/A residues in helix packing and closed-state stability supported by functional data? Answers to these questions will help to understand the interactions of residues in the bundle crossing region and help to clarify the specific impact of pore residues on activation and inactivation.

---

[*] The work was supported by Fonds zur Förderung der Wissenschaftlichen Forschung (Austrian Science Fund) Grants P22600 (to S. B.), W1232 (to S. H.), and P19614 (to S. H.).
[S] The on-line version of this article (available at http://www.jbc.org) contains supplemental Figs. S1-S5.
[✗] *Author's Choice*—Final version full access.
[1] These authors contributed equally to this work.
[2] To whom correspondence should be addressed. Tel.: 43-14277-55310; Fax: 43-14277-9553; E-mail: steffen.hering@univie.ac.at.
[3] The abbreviation used is: TS, Timothy syndrome.

Link between Activation and Inactivation

EXPERIMENTAL PROCEDURES

Conservation Analysis—Sequences of voltage-gated calcium channels (13) were downloaded from the International Union of Basic and Clinical Pharmacology data base (14). Sequences of all channels from *Homo sapiens* were used to search NCBI BLAST (blastp) (15). Initial sequence alignment was performed using ClustalW (16). Sequences of selected potassium channel crystal structures (KcsA, MthK, K_vAP, $K_v1.2$, and MlotiK) were aligned manually with Ca_V channels using GENEDOC (17).

Homology Modeling—The homology model of the open $Ca_V1.2$ channels, based on the K_vAP crystal structure and a NaChBac model, incorporating an insertion at the position of the conserved asparagines, was published previously (8). A model of the closed conformation using the same alignment as for the open conformation (indel at Asn) was generated using Modeller9v7 (18). The KcsA crystal structure and the NaChBac model were used as templates. Coordinates of the model can be found in the supplemental material.

Accessibility Calculations—Amino acid accessibilities were calculated using the WHAT IF webserver (19). The accessible surface area in this program is defined as the area at the Van der Waals surface that is accessible by a water molecule (1.4 Å). The units of accessible surface are $Å^2$. Re-entrant surfaces are not considered by WHAT IF.

Mutagenesis—The $Ca_V1.2$ $α_1$-subunit coding sequence (GenBank™ X15539) in-frame 3′ to the coding region of a modified green fluorescent protein was kindly donated by Dr. M. Grabner (20). Substitutions in segment IS6, IIS6, and IVS6 of the $Ca_V1.2$ $α_1$-subunit were introduced using the QuikChange® Lightning site-directed mutagenesis kit (Stratagene) with mutagenic primers according to the manufacturer's instructions. Mutations were introduced in segment IS6 in position 432 (G432A/S/V/N/M/W), in helix IIS6 in position 780 (A780G/N/P/T/V/W), in IVS6 in position 1503 (A1503G/P/M/N/T/V/W), and one double mutation in IS6 (G432S/S435G). All constructs were checked by restriction site mapping and sequencing.

Cell Culture and Transient Transfection—Human embryonic kidney tsA-201 cells were grown at 5% CO_2 and 37 °C to 80% confluence in Dulbecco's modified Eagle's/F-12 medium supplemented with 10% (v/v) fetal calf serum and 100 units/ml penicillin/streptomycin. Cells were split with trypsin/EDTA and plated on 35-mm Petri dishes (Falcon) at 30–50% confluence ~16 h before transfection. Subsequently tsA-201 cells were co-transfected with cDNAs encoding wild-type or mutant $Ca_V1.2$ $α_1$-subunits with auxiliary $β_{2a}$ and $β_{2c}$ (21) as well as $α_2$-$δ_1$ (22) subunits. The transfection of tsA-201 cells was performed using the FuGENE 6 transfection reagent (Roche Applied Science) following standard protocols. tsA-201 cells were used until passage number 15. No variation in channel gating related to different cell passage numbers was observed.

Ionic Current Recordings and Data Acquisition—Barium currents (I_{Ba}) through voltage-gated Ca^{2+} channels were recorded at 22–25 °C by patch-clamping (23) using an Axopatch 200A patch clamp amplifier (Axon Instruments, Foster City) 36–48 h after transfection. The extracellular bath solution (5 mM $BaCl_2$, 1 mM $MgCl_2$, 10 mM HEPES, 140 mM choline chloride) was titrated to pH 7.4 with methanesulfonic acid. Patch pipettes with resistances of 1–4 megaohms were made from borosilicate glass (Clark Electromedical Instruments) and filled with pipette solution (145 mM CsCl, 3 mM $MgCl_2$, 10 mM HEPES, 10 mM EGTA 10), titrated to pH 7.25 with CsOH. All data were digitized using a DIGIDATA 1200 interface (Axon Instruments), smoothed by means of a four-pole Bessel filter, and saved to disc. 100-ms current traces were sampled at 10 kHz and filtered at 5 kHz; tail currents were sampled at 50 kHz and filtered at 10 kHz. Leak currents were subtracted digitally using average values of scaled leakage currents elicited by a 10-mV hyperpolarizing pulse or electronically by means of an Axopatch 200 amplifier (Axon Instruments). Series resistance and offset voltage were routinely compensated for. The pClamp software package (Version 7.0 Axon Instruments) was used for data acquisition and preliminary analysis. Microcal Origin 7.0 was used for analysis and curve-fitting.

The voltage dependence of activation was determined from I-V curves and fitted to

$$m_\infty = \frac{1}{1 + \exp\frac{V_{0.5,act} - V}{k_{act}}} \qquad (\text{Eq. 1})$$

The time course of current activation was fitted to a monoexponential function, $I(t) = A \cdot \exp(1/τ) + C$, where $I(t)$ is current at time t, A is the amplitude coefficient, $τ$ is the time constant, and C is the steady-state current. Data are given as the mean ± S.E.

Calculation of Amino Acid Descriptors—Physicochemical descriptors were calculated for amino acids using the Molecular Operating Environment (MOE; Version 2008.10, Chemical Computing Group, Inc., Montreal, QC, Canada) (24).

RESULTS

The "Timothy Gly-432" Is Part of a Motif of Small Residues at the Inner Channel Mouth—To analyze the degree of conservation of the TS Gly-402, a sequence alignment was performed. Sequences of the 10 different voltage-gated calcium channels genes encoded in the human genome were downloaded from the International Union of Basic and Clinical Pharmacology data base. The human sequences were used to search NCBI BLAST using the blastp program. The pore-forming segments of Ca_V (S5, P, S6) from different species were aligned using standard parameters in ClustalW. Fig. 1 shows the multiple sequence alignment of selected Ca_V channels, and the complete alignment of all human S6 segments can be found in supplemental Fig. S1. Fig. 1 shows that Gly-432 is part of a motif of small amino acids (Gly-432(IS6)-Ala-780(IIS6)-Gly-1193(IIIS6)-Ala-1503(IVS6)) at the inner channel mouth. The conservation analysis reveals that these positions either have a glycine or an alanine. The only exceptions are T-type channels, which have a valine at this position in IS6. We will refer to this structural motif of Ca_V as G/A/G/A.

Mutations of Gly-432 Shift the Activation Curve—In the absence of experimentally determined structures of calcium channels, homology models have been used successfully to explain experimental data and to suggest new experiments (25–

Link between Activation and Inactivation

```
Cav1.2   I      DWPWIYFVTLIIIGSFFVLNLVLGVLSGEFS      379-409
Cav1.2   I(r)   ELPWVYFVSLVIFGSFFVLNLVLGVVLSGEFS     409-439
Cav2.1   I      TWNWLYFIPLIIIGSFFMLNLVLGVLSGEFA      334-364
Cav3.1   I      FYNFIYFILLIIVGSFFMINLCLVVIATQFS      369-399
Cav1.2   II     MLVCIYFIILFICGNYILLNVFLAIAVDNLA      727-757
Cav2.1   II     MVFSIYFIVLTLFGNYTLLNVFLAIAVDNLA      688-718
Cav3.1   II     SWAALYFIALMTFGNYVLFNLLVAILVEGFQ      938-968
Cav1.2   III    VEISIFFIIYIIIIAFFMMNIFVGFVIVTFQ      1160-1190
Cav2.1   III    MEMSIFYVVYFVVFPFFFVNIFVALIIITFQ      1485-1515
Cav3.1   III    PWMLLYFISFLLIVAFFVLNMFVGVVVENFH      1511-1541
Cav1.2   IV     SFAVFYFISFYMLCAFLIINLFVAVIMDNFD      1498-1528
Cav2.1   IV     EFAYFYFVSFIFLCSFLMLNLFVAVIMDNFE      1785-1815
Cav3.1   IV     VISPIYFVSFVLTAQFVLVNVVIAVLMKHIE      1825-1855
KcsA            LWGRLVAVVVMVAGITSFG-LVTAALATWFV       86-115
MlotiK          FAGRVLAGAVMMSGIGIFG-LWAGILATGFY      186-216
Kv1.2           IGGKIVGSLCAIAGVLTIA-LPVPVIVSNFN      385-414
KvAP            PIGEVIGIAVMLTGISALT-LLIGTVSNMFQ      207-236
```

FIGURE 1. **Multiple sequence alignment of S6 helices of Ca$_V$ channels with selected K$^+$ channels.** Conserved G/A/G/A residues are shaded green, and interacting phenylalanines (neighboring domains) are shaded blue. The Timothy glycine is underlined green, and the PVP motif of K$_v$1.2 is underlined black.

29). Our modeling data on Ca$_V$1.2 suggest that G/A/G/A plays an essential role in helix packing. In light of this hypothesis, we investigated whether replacement of Gly-432 by residues with different properties (e.g. size and hydrophobicity) would affect the voltage dependence of channel activation. Fig. 2, A and B, illustrate the corresponding effects on the steady-state activation curves. Replacement of Gly-432 by methionine, asparagine, and tryptophan shifts channel activation toward hyperpolarization. Interestingly, G432A and G432V shifted the curve to the right, suggesting a stabilization of the closed or a destabilization of the open states. Changes in activation and deactivation gating for selected mutants are illustrated by current traces shown in Fig. 2, C and D. The slowing of channel activation and deactivation (voltage dependences of activation and deactivation time constants) for G432N accompanying the leftward shift of the activation curve and fast activation/deactivation kinetics of G432A inducing a rightward shift are exemplified (Fig. 2, E and F).

Substitutions of Gly-432 Remove Channel Inactivation—In contrast to the differing effects on channel activation (Fig. 2, Table 1), all mutations of Gly-432 tested reduced inactivation regardless of the properties of the new side chain. This is evident from current traces under long-lasting depolarizations for all mutants (Fig. 3). The changes in the steady-state inactivation curves for mutation to alanine (G432A) and asparagines (G432N) are exemplified in Fig. 3A. The remaining currents after 3000-ms depolarizations to the peak current voltage of the current voltage activation curves are given in Table 1.

Mutations of Ala-780, Gly-1193, and Ala-1503 (Homologous to Gly-432) Affect Channel Activation but Preserve Inactivation—None of the mutations in positions Ala-780, Gly-1193, and Ala-1503 slowed channel inactivation, as observed with substitutions of Gly-432. The modulatory effects of mutations in these positions are illustrated in Fig. 4. We observed a strong correlation between the shifts of the activation and inactivation curves (Fig. 5, $r = 0.91$, $p < 0.001$) in domains II, III, and IV (see also Table 1), suggesting that the coupling between the two processes was intact. These data indicate a special role of Gly-432 in inactivation gating, in that only mutations of Gly-432

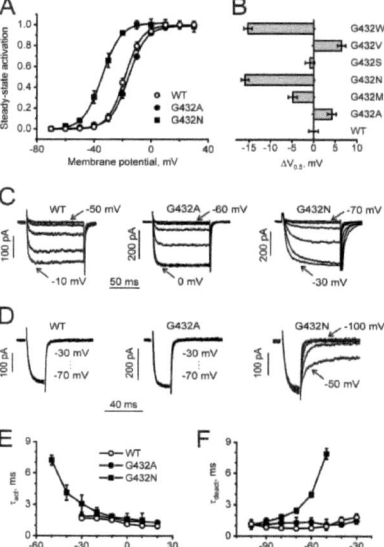

FIGURE 2. **Gating changes induced by mutations of Gly-432 in segment IS6 of Ca$_V$1.2.** A, averaged activation curves of wild-type ($n = 9$) and mutants G432A ($n = 7$) and G432N ($n = 7$) are shown. B, shown is a shift of activation curves induced by substitutions of Gly-432 by residues of different physicochemical properties. C, representative families of I_{Ba} through wild-type and the indicated mutant channels during depolarizing test pulses from -100 mV in 10-mV increments (first and last potentials are indicated) are shown. D, representative tail currents of WT and the indicated mutant channels are shown. Currents were activated during a 20-ms conditioning depolarization to -10 mV for wild type, 0 mV for G432A, and -20 mV for G432N. Deactivation was recorded during subsequent repolarizations in 10-mV increments starting from -100 mV. E, mean time constants of channel activation of WT ($n = 9$) and mutants G432A ($n = 7$) and G432N ($n = 7$) are plotted against test potential. F, mean time constants of channel deactivation of wild type ($n = 8$) and mutants G432A ($n = 6$) and G432N ($n = 6$) are plotted against test potential. Time constants were estimated by fitting current activation and deactivation to a mono-exponential function.

diminished inactivation, whereas substitutions in homologous positions in other segments either had no effect on inactivation or even made it faster.

Structural Analysis of the G/A/G/A Motif—To visualize the location of G/A/G/A, homology models of Ca$_V$1.2 in open and closed conformation were analyzed. The open channel conformation has been published previously (30). A model of the closed Ca$_V$1.2 channel, including an insertion at the highly conserved asparagines, was developed by basing the structures of P-helices and selectivity filter segments on the NaChBac model (31), as both channels possess the

Link between Activation and Inactivation

TABLE 1
Pore mutations affect voltage-dependent gating of $Ca_V 1.2$
Midpoints, slope factors of the activation, inactivation curves, and fractions of non-inactivated channels after a 3-s depolarization to the peak current potential of the current voltage relationship (I-V curve) are shown. Numbers of experiments are indicated in parentheses.

Construct	$V_{0.5}$	k_{act}	$V_{0.5,inact}$	k_{inact}	r_{3000}
	mV	mV	mV	mV	
WT	-18.4 ± 0.8 (9)	6.0 ± 0.7	-41.5 ± 0.9 (5)	6.4 ± 0.9	0.35 ± 0.03
IS6 G432[a]					
G432A	-14.2 ± 0.6 (7)	6.2 ± 0.5			0.77 ± 0.05
G432M	-23.2 ± 0.6 (7)	7.7 ± 0.6			0.68 ± 0.07
G432N	-34.4 ± 0.4 (7)	6.5 ± 0.5			0.72 ± 0.04
G432S	-19.2 ± 0.7 (7)	5.9 ± 0.7			0.74 ± 0.07
G432V	-12.0 ± 0.6 (6)	5.7 ± 0.5			0.86 ± 0.04
G432W	-33.7 ± 0.6 (8)	4.7 ± 0.4			0.58 ± 0.04
IIS6 A780					
A780G	-29.4 ± 0.6 (5)	6.2 ± 0.6	-54.4 ± 1.3 (4)	6.2 ± 1.1	0.11 ± 0.06
A780N	-43.8 ± 1.0 (7)	6.8 ± 0.9	-67.4 ± 0.7 (4)	4.6 ± 0.9	0.25 ± 0.04
A780P	-45.0 ± 0.7 (5)	5.9 ± 0.8	-39.4 ± 1.2 (4)	5.2 ± 1.9	0.34 ± 0.07
A780T	-44.0 ± 0.6 (5)	5.1 ± 0.6	-59.8 ± 0.9 (4)	7.5 ± 0.9	0.24 ± 0.04
A780V	-34.6 ± 0.8 (8)	6.6 ± 0.7	-54.3 ± 0.8 (4)	4.8 ± 0.8	0.10 ± 0.05
A780W	-49.3 ± 0.6 (7)	5.9 ± 0.5	-68.0 ± 0.8 (4)	6.3 ± 0.9	0.16 ± 0.05
IIIS6 G1193[b]					
G1193A	-12.1 ± 1.0 (6)	5.7 ± 0.7	-25.2 ± 1.1 (6)	8.8 ± 0.9	0.27 ± 0.07
G1193M	-33.6 ± 0.7 (8)	6.4 ± 0.6	-52.8 ± 1.0 (4)	6.1 ± 0.9	0.34 ± 0.08
G1193N	-46.5 ± 1.0 (5)	5.6 ± 0.8	-61.3 ± 0.9 (4)	6.9 ± 0.9	0.38 ± 0.08
G1193P	-37.7 ± 0.7 (7)	6.3 ± 0.6	-64.6 ± 1.0 (4)	6.1 ± 0.8	0.18 ± 0.07
G1193Q	-39.2 ± 0.8 (5)	7.0 ± 0.7	-64.4 ± 1.2 (4)	8.4 ± 1.1	0.29 ± 0.08
G1193T	-49.8 ± 1.0 (7)	5.1 ± 0.6	-70.0 ± 1.1 (5)	5.4 ± 1.1	0.16 ± 0.07
G1193V	-38.1 ± 0.7 (8)	5.6 ± 0.6	-60.4 ± 0.8 (5)	7.5 ± 0.6	0.14 ± 0.08
IVS6 A1503					
A1503G	-41.4 ± 0.8 (5)	5.8 ± 0.5	-54.0 ± 1.0 (4)	4.9 ± 0.8	0.31 ± 0.07
A1503M	-29.2 ± 0.7 (8)	5.9 ± 0.7	-60.1 ± 1.0 (4)	5.5 ± 1.0	0.26 ± 0.08
A1503N	-31.5 ± 0.7 (5)	6.3 ± 0.6	-45.1 ± 1.1 (4)	6.0 ± 1.0	0.32 ± 0.09
A1503T	-17.9 ± 0.8 (7)	8.7 ± 0.8	-38.0 ± 1.2 (4)	6.9 ± 1.1	0.29 ± 0.09
A1503V	-11.4 ± 0.8 (5)	7.2 ± 0.6	-37.2 ± 1.2 (4)	6.8 ± 1.2	0.24 ± 0.06
A1503W	-16.9 ± 0.6 (9)	6.6 ± 0.6	-38.9 ± 1.3 (4)	7.5 ± 0.8	0.33 ± 0.09
A1503P	no current				
Double mutation					
G432S/S435G	-19.8 ± 0.9 (5)	6.3 ± 0.7	-39.2 ± 1.2 (4)	5.7 ± 1.1	0.54 ± 0.07

[a] $V_{0.5,inact}$ lack of inactivation prevented quantitative estimations.
[b] $V_{0.5,act}$ data are from Beyl et al. (24).

same selectivity signature sequence. The S5 and S6 segments are based on the KcsA crystal structure. The alignment is shown in Fig. 1.

Residues in closed K^+ channel crystal structures of KcsA and MlotiK that align with the G/A/G/A motif are positioned at the M2 helix-helix interface (Fig. 6, A and B). These helices correspond to S6 helices in calcium channels. The small amino acids in positions homologous to Ca_V G/A/G/A motif interact with bulky hydrophobic residues from the adjacent M2 helices. Interestingly, in both crystal structures the small Ala-108/Gly-208 are in close contact with a conserved phenylalanine (see alignment, Fig. 1) located two turns below in the neighboring pore helix. Our homology model of the closed conformation suggests similar interactions in $Ca_V 1.2$ (Fig. 6, C–F). Small amino acids Gly-432, Ala-780, Gly-1193, and Ala-1503 are located at the S6-S6 interface, tightly surrounded by hydrophobic residues from the adjacent S6 segments. A structural hypothesis (Fig. 6, C–F) illustrates putative interactions of the G/A/G/A residues with adjacent bulky residues, e.g. phenylalanines in domains I, III, and IV and leucine in domain II. All residues that are located within a short distance (<7 Å of Cβ atom) of G/A/G/A according to our homology model are given in Table 2.

To further characterize this putative packing motif, the accessibility of the surrounding residues was calculated with the WHAT IF online server (19) (see Table 3). Calculations using KcsA and MlotiK crystal structures show that the amino acids corresponding to the G/A/G/A motif are inaccessible. Similar results are obtained when calculating the accessible molecular surface in our closed $Ca_V 1.2$ homology model. The conserved phenylalanines are also generally inaccessible, with the exception of Phe-114 from the KcsA crystal structure. This might be due to incompleteness of the crystal structure, which lacks the slide helices.

A comparison of closed and open crystal structure conformations of the bacterial potassium channel KcsA reveals large conformational changes upon gating, resulting in a displacement of this amino acid by more than 7 Å. Accessibility calculations show that this residue is accessible to water in the open channel conformation. Because several open and open-inactivated crystal structures are available, we calculated values for three different KcsA crystal structures (32) with the PDB identifiers 3f5w, 3fb7, and 3f7v. The obtained accessibility values (range 9.9–17.2 Å²) might reflect different degrees of channel opening. Similar accessibilities are obtained when calculating the accessible molecular surface in our open $Ca_V 1.2$ homology model. The values are 14.4 Å² (Gly-432), 11.8 Å² (Ala-780), 11.1 Å² (G1193), and 18.5 Å² (A1503).

DISCUSSION

Reduced inactivation by TS mutations G406R and G402S results in sustained membrane depolarization and enhanced

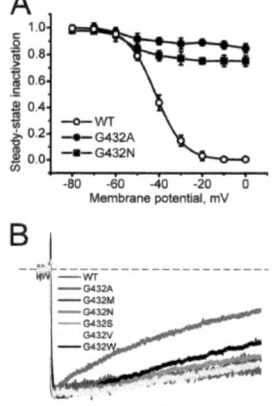

FIGURE 3. **Inactivation changes induced by mutations of Gly-432 in segment IS6 of Ca$_V$1.2.** *A*, averaged inactivation curves of wild-type ($n = 5$) and mutants G432A ($n = 4$) and G432N ($n = 4$). *B*, representative I_{Ba} through wild-type and indicated mutant channels during depolarizing test pulses from -100 mV to the peak potential of the I-V curve.

calcium entry through L-type channels into neuronal, myocardial, and other cell types (2). This slowing of channel inactivation is the key to understanding the wide spectrum of developmental abnormalities and pathological changes in different organs leading to cardiac arrhythmias, autism spectrum disorders, and other diseases (2). TS mutations affect the mechanism of calcium-dependent inactivation that occurs irrespective of expression of a fast (β_{1c}-) or slow β_{2a}-subunit (3, 4). Later studies revealed that mutation G436R (corresponding to the TS mutation G406R in humans) slows channel activation and deactivation, which may augment the arrhythmogenic effects of this mutation (33). Splawski *et al.* (2) emphasize that substitution of Gly-406 by different amino acids reduces inactivation similar to G406R. This finding was supported by data of Raybaud *et al.* (3) who substituted 11 different residues for the corresponding Gly-436 in the rabbit channel and found that all these mutations reduced voltage-dependent inactivation. However, the role of these residues in stabilization of the different conformational states of Ca$_V$1.2 remained elusive.

Splawski *et al.* (2) hypothesized, in analogy to potassium channels, that Gly-406 and Gly-402 may act as the hinge points during channel activation. Raybaud *et al.* (3) discussed a potential role of G402S and G406R in helix packing. In continuation of the work by Splawski *et al.* (2) and Raybaud *et al.* (3), we have substituted Gly-432 by residues of different physicochemical properties. Our homology model suggests that the Timothy Gly-432 is part of a motif of tightly packed small amino acids

(Fig. 6, *A–F*). It was, therefore, interesting to compare gating changes induced by mutations in these positions of all four S6 segments.

Mutations of G/A/G/A Residues Affect Activation Gating— All mutations of the TS Gly-432 and homologous Ala-780, Gly-1193, and Ala-1503 affected the stability of the channel pore. The majority of mutations induced leftward shifts of the activation curves that were always accompanied by a slowing down of channel deactivation, suggesting a stabilization of the open and destabilization of the closed state (9, 24). Four mutations, G432A, G432V, G1193A, and A1503V, induced rightward shifts of the steady-state curve and exhibited corresponding fast activation and deactivation (Figs. 2 and 4), indicating a stabilization of the closed channel state.

To test the structural hypothesis that size is an important determinant of pore stability (supplemental Fig. S2) in the G/A/G/A positions, we applied mutation-correlation analysis (24). In all four positions the insertion of larger residues disturbed pore stability, as evident from shifts of the activation curve. In line with these data, characteristics of amino acid size (either molecular weight or Van der Waals volume) correlated in combination with hydrophobicity or flexibility indices with the shifts of the activation curve (supplemental Fig. S2). Despite their common impact on activation gating (all residues of G/A/G/A contribute to closed state stability), these residues interact in a domain-specific manner with their surrounding residues. As illustrated in supplemental Fig. S2, descriptors characterizing the size of the substituting residues (either molecular weight in domains I-III or Van der Waals volume in domain IV) correlate with the shifts of the activation curves. Domain-specific (asymmetric) contributions are evident from different impacts of hydrophobicity indices (domains I-III) and side chain flexibility (domain IV). Interestingly, mutating a small and flexible Gly-432 to a small but more rigid alanine or valine in domain I stabilized the closed state by shifting the activation curve by 5–7 mV. In domain IV, replacing a rigid Ala-1503 by a flexible glycine destabilizes the closed state by shifting the activation curve by ≈ -20 mV. The asymmetric impact of pore mutations (*i.e.* domain specific contributions of G/A/G/A substitutions) on Ca$_V$1.2 gating warrants further research.

Structural Model of the G/A/G/A Region—To illustrate the structural location of the TS Gly-432 and the G/A/G/A motif residues, we refer to crystal structures of potassium channels and designed a homology model. The model supports the idea that the small amino acids are in close contact with bulky hydrophobic residues from neighboring S6 segments, thereby providing domain-domain interfaces (Fig. 6). In other words, these homologous residues are likely to be involved in tight S6-S6 packing of the closed channel conformation. If we see these interactions as stabilizing the closed conformation, then the insertion of larger residues would be expected to disrupt the packing and destabilize the closed state, as illustrated in supplemental Fig. S3.

Exclusive Role of Gly-432 in Channel Inactivation—In this study we asked the question of how structural changes in Gly-432 and homologous residues Ala-780, Gly-1193, and Ala-1503

Link between Activation and Inactivation

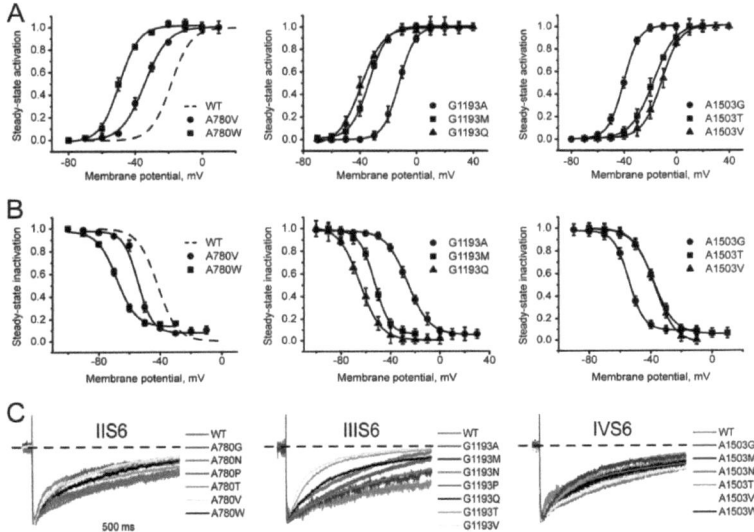

FIGURE 4. **Mutations in the pore segments IIS6, IIIS6, and IVS6 in positions homologous to Gly-432 in IS6 induce significant shifts of activation and inactivation curves.** *A* and *B*, shown is the averaged activation (*A*) and inactivation (*B*) curves of wild type and mutations of Ala-780 in segment IIS6 (*left*), Gly-1193 in segment IIIS6 (*middle*), and Ala-1503 in segment IVS6 (*right*). All mutations are summarized in Table 1. *C*, shown are representative I_{Ba} through wild type and the indicated mutant channels in segments IIS6-IVS6 during depolarizing test pulses from −100 mV to the peak potentials of the current-voltage relationships. No deceleration of inactivation was observed (compare with Gly-432 in IS6 in Fig. 3).

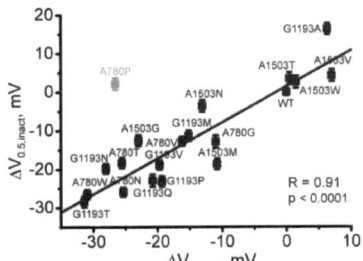

FIGURE 5. **Correlation between shifts of the midpoint of the inactivation and activation curves for G/A/G/A mutations in segments IIS6-IVS6.** A proline mutation of Ala-780 (*shaded*) was excluded from correlation analysis as this residue is likely to disturb the link between voltage sensor and the channel pore (9, 24).

affect channel inactivation. Our data revealed a high degree of positional specificity for mutations in position Gly-432. All mutations of Gly-432 slowed down or prevented inactivation (Fig. 3, Table 1), whereas substitutions in domains II-IV did either not affect, or even accelerated inactivation (Fig. 4*C*). Parallel shifts of the activation and inactivation curves (Fig. 5) indicate that mutations in these three domains do not affect the coupling between activation and inactivation supporting a gating scheme (R)est→(O)pen→(I)nactivated.

However, mutations in position Gly-432 suppress inactivation regardless of the stability of the activation gates. Any mutation that either stabilizes (*e.g.* G432A/V) or destabilizes (G432S/M/N/W) the closed resting state prevents channel inactivation, supporting a gating scheme R → O ↛ I. This is evident from Figs. 2 and 3, illustrating the observed stabilization of the closed state (shifts of activation curve to the *right*) and destabilization of the closed state (different shifts of the curve to the *left* or to the *right*, see supplemental Fig. S2A). Thus, despite the similar impact of Gly-432 and other residues of the G/A/G/A ring in channel activation, only Gly-432 (IS6) is essential for inactivation in an "all or none" manner.

Conclusions and Outlook—Our structure-activity studies suggest that the TS Gly-432 in domain I of $Ca_V 1.2$ forms part of a ring of small and tightly packed residues (G/A/G/A) interacting with adjacent bulky hydrophobic amino acids. The identification of the particular interaction partners of the G/A/G/A ring in adjacent segments remains a challenge

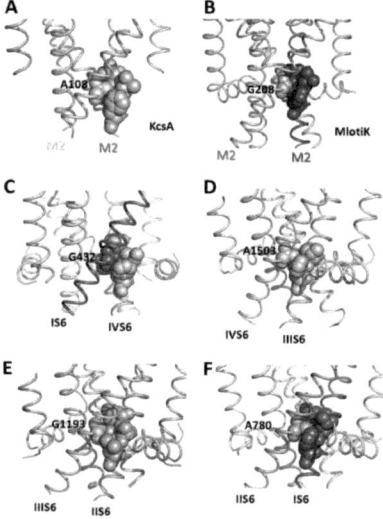

FIGURE 6. **Hypothetical role of small amino acids in helix-helix packing of voltage-gated ion channels.** Shown are the location of an alanine (A) in the closed conformation crystal structures of KcsA and the location of a corresponding glycine (B) in Mlotik (34). C–F illustrate the location of the G/A/G/A residues interacting with bulky hydrophobic residues from the neighboring S6 segments (shown as spheres).

for future experimental studies (see Table 3). These small residues in the lower third of the $Ca_V1.2$ S6 segments are homologous to a ring of small residues in MlotiK and KcsA crystal structures. Systematic mutational studies and correlation analysis of Ca_V revealed that hydrophobic interactions in this region stabilize the channel closed state. The small size is apparently required for tight-fitting interactions with neighboring bulky residues of $Ca_V1.2$. The apparent structural homologies between KcsA and MlotiK and Ca_V in the pore region provide a guideline for the alignment of S6 segments. Only mutations of the TS Gly-432 disrupted the link between activation and inactivation, whereas substitution of the other alanines or the IIIS6 glycine in most cases simultaneously shifted activation and inactivation. The mechanism behind the unique effect of Gly-432 mutations on inactivation remains to be elucidated.

Ca^{2+} channels are expressed as a single polypeptide assembling into four pseudo-subunits. Although the pseudo-4-fold symmetry might be maintained in S6 helices, it is not preserved in the intracellular domains, where sequence length and interaction partners vary.

It is tempting to speculate that flexibility in IS6 is required for proper transmission of β-subunit-mediated effects on inactivation. Only Gly-432 in segment IS6 is linked to the β interaction domain (BID) located on I-II linker (35, 36). Mutations in position Gly-432 disrupt inactivation irrespective whether the $Ca_V1.2$ comprises a "fast-inactivating" β_1 or a "slow-inactivating" β_{2a} subunit (exemplified in supplemental Fig. S4), suggesting that the mutated $Ca_V1.2$ α-subunits (G432X) interact with their β subunits. Similar findings were previously reported for the Timothy mutations G406R (4). We have therefore substituted a flexible glycine for a serine one turn below Gly-432. The introduction of helix flexibility next to the TS mutation G432S partially rescued channel inactivation (supplemental Fig. S5). Flexibility in the lower third of segment IS6 may thus be a molecular determinant of β-subunit-mediated inactivation.

TABLE 2
Potential interactions of G/A/G/A residues in the closed channel state (see the model in Fig. 6)

Member of G/A/G/A	Neighboring residues (β–carbons <7 Å apart)
Gly-432 (IS6)	IS6: Asn-428, Leu-429, Leu-431, Val-433, Ser-435, Gly-436, IVS6: Val-1502, Ile-1505, Met-1506, Phe-1509
Ala-780 (IIS6)	IS6: Leu-431, Gly-432, Leu-434, Ser-435, Gly-436, Phe-438
	IIS6: Asn-776, Val-777, Leu-779, Ile-781, Val-783, Asp-784
Gly-1193 (IIIS6)	IIS6: Leu-779, Ala-782, Val-783, Leu-786
	IIIS6: Asn-1189, Ile-1190, Val-1192, Phe-1194, Val-1195, Ile-1196, Val-1197
	IVS6: Asn-1499, Ala-1503
Ala-1503 (IVS6)	IIIS6: Gly-1193, Val-1195, Ile-1196, Phe-1199
	IVS5: Lys-1385
	IVS6: Asn-1499, Leu-1500, Val-1502, Val-1504, Met-1506, Asp-1507

TABLE 3
Accessibility of residues at the contact interface between S6 (M2) helices
The G/A/G/A motif is shown in bold; conserved phenylalanines are underlined.

Channel	Accessibilities of Amino Acids in the Bundle-crossing Region (Å²)
KcsA (PDB code 1k4c)	Thr-107 (5.9); **Ala-108 (0.17)**; Ala-109 (14.19); Leu-110 (10.1), Ala-111 (2.44), Thr-112 (11.5), Trp-113 (56.8), Phe-114 (20.2)
MlotiK (PDB code 3beh)	Ala-207 (1.7); **Gly-208 (0)**; Ile-209 (0.1); Leu-210 (1.2), Ala-211 (4.1), Thr-212 (2.4); Gly-213 (0); Phe-114 (1.9)
$Ca_V1.2$ I	Leu-431 (0); **Gly-432 (0)**; Val-433 (0.6); Leu-434 (1.2); Ser-435 (0), Gly-436 (0); Glu-437 (17.1); Phe-438 (1.5)
$Ca_V1.2$ II	Leu-779 (0.3); **Ala-780 (0)**; Ile-781 (7.3); Ala-782 (1.7); Val-783 (1.5); Asp-784 (1.5); Asn-785 (21.5); Leu-786 (2.9)
$Ca_V1.2$ III	Val-1192 (0.1); **Gly-1193 (0)**; Phe-1194 (13.9); Val-1195 (6.9); Ile-1196 (0.1); Val-1197 (19.8); Thr-1198 (19.8); Phe-1199 (3.4)
$Ca_V1.2$ IV	Val-1502 (0); **Ala-1503 (0)**; Val-1504 (5.5); Ile-1505 (9.7); Met-1506 (0.3); Asp-1507 (1.0); Asn-1508 (17.6); Phe-1509 (1.6)

Acknowledgment—We thank Waheed Shabbir for technical assistance.

REFERENCES

1. Splawski, I., Timothy, K. W., Sharpe, L. M., Decher, N., Kumar, P., Bloise, R., Napolitano, C., Schwartz, P. J., Joseph, R. M., Condouris, K., Tager-Flusberg, H., Priori, S. G., Sanguinetti, M. C., and Keating, M. T. (2004) *Cell* **119**, 19–31
2. Splawski, I., Timothy, K. W., Decher, N., Kumar, P., Sachse, F. B., Beggs, A. H., Sanguinetti, M. C., and Keating, M. T. (2005) *Proc. Natl. Acad. Sci. U.S.A.* **102**, 8086–8096; discussion 8086–8088
3. Raybaud, A., Dodier, Y., Bissonnette, P., Simoes, M., Bichet, D. G., Sauvé, R., and Parent, L. (2006) *J. Biol. Chem.* **281**, 39424–39436
4. Barrett, C. F., and Tsien, R. W. (2008) *Proc. Natl. Acad. Sci. U.S.A.* **105**, 2157–2162
5. Shi, C., and Soldatov, N. M. (2002) *J. Biol. Chem.* **277**, 6813–6821
6. Hering, S., Berjukow, S., Sokolov, S., Marksteiner, R., Weiss, R. G., Kraus, R., and Timin, E. N. (2000) *J. Physiol.* **528**, 237–249
7. Stotz, S. C., and Zamponi, G. W. (2001) *Trends Neurosci.* **24**, 176–181
8. Kudrnac, M., Beyl, S., Hohaus, A., Stary, A., Peterbauer, T., Timin, E., and Hering, S. (2009) *J. Biol. Chem.* **284**, 12276–12284
9. Hohaus, A., Beyl, S., Kudrnac, M., Berjukow, S., Timin, E. N., Marksteiner, R., Maw, M. A., and Hering, S. (2005) *J. Biol. Chem.* **280**, 38471–38477
10. Hardman, R. M., Stansfeld, P. J., Dalibalta, S., Sutcliffe, M. J., and Mitcheson, J. S. (2007) *J. Biol. Chem.* **282**, 31972–31981
11. Irizarry, S. N., Kutluay, E., Drews, G., Hart, S. J., and Heginbotham, L. (2002) *Biochemistry* **41**, 13653–13662
12. Paynter, J. J., Sarkies, P., Andres-Enguix, I., and Tucker, S. J. (2008) *Channels* **2**, 413–418
13. Catterall, W. A., Perez-Reyes, E., Snutch, T. P., and Striessnig, J. (2005) *Pharmacol. Rev.* **57**, 411–425
14. Harmar, A. J., Hills, R. A., Rosser, E. M., Jones, M., Buneman, O. P., Dunbar, D. R., Greenhill, S. D., Hale, V. A., Sharman, J. L., Bonner, T. I., Catterall, W. A., Davenport, A. P., Delagrange, P., Dollery, C. T., Foord, S. M., Gutman, G. A., Laudet, V., Neubig, R. R., Ohlstein, E. H., Olsen, R. W., Peters, J., Pin, J. P., Ruffolo, R. R., Searls, D. B., Wright, M. W., and Spedding, M. (2009) *Nucleic Acids Res.* **37**, D680–D685
15. Altschul, S. F., Gish, W., Miller, W., Myers, E. W., and Lipman, D. J. (1990) *J. Mol. Biol.* **215**, 403–410
16. Chenna, R., Sugawara, H., Koike, T., Lopez, R., Gibson, T. J., Higgins, D. G., and Thompson, J. D. (2003) *Nucleic Acids Res.* **31**, 3497–3500
17. Nicholas, K. B., Nicholas, H. B., Jr., and Deerfield, D. W., II (1997) *EMBNEW NEWS* **4**, 14
18. Eswar, N., Webb, B., Marti-Renom, M. A., Madhusudhan, M. S., Eramian, D., Shen, M. Y., Pieper, U., and Sali, A. (2007) *Curr. Protoc. Protein Sci.*, Chapter 2, Unit 2.9
19. Vriend, G. (1990) *J. Mol. Graph* **8**, 52–56, 29
20. Grabner, M., Dirksen, R. T., and Beam, K. G. (1998) *Proc. Natl. Acad. Sci. U.S.A.* **95**, 1903–1908
21. Perez-Reyes, E., Castellano, A., Kim, H. S., Bertrand, P., Baggstrom, E., Lacerda, A. E., Wei, X. Y., and Birnbaumer, L. (1992) *J. Biol. Chem.* **267**, 1792–1797
22. Ellis, S. B., Williams, M. E., Ways, N. R., Brenner, R., Sharp, A. H., Leung, A. T., Campbell, K. P., McKenna, E., Koch, W. J., and Hui, A. (1988) *Science* **241**, 1661–1664
23. Hamill, O. P., Marty, A., Neher, E., Sakmann, B., and Sigworth, F. J. (1981) *Pflugers Arch.* **391**, 85–100
24. Beyl, S., Depil, K., Hohaus, A., Stary-Weinzinger, A., Timin, E., Shabbir, W., Kudrnac, M., and Hering, S. (2011) *Pflugers Arch.* **461**, 53–63
25. Huber, I., Wappl, E., Herzog, A., Mitterdorfer, J., Glossmann, H., Langer, T., and Striessnig, J. (2000) *Biochem. J.* **347**, 829–836
26. Stary, A., Kudrnac, M., Beyl, S., Hohaus, A., Timin, E. N., Wolschann, P., Guy, H. R., and Hering, S. (2008) *Channels* **2**, 216–223
27. Bruhova, I., and Zhorov, B. S. (2010) *J. Gen. Physiol.* **135**, 261–274
28. Tikhonov, D. B., and Zhorov, B. S. (2011) *J. Biol. Chem.* **286**, 2998–3006
29. Lipkind, G. M., and Fozzard, H. A. (2001) *Biochemistry* **40**, 6786–6794
30. Stary, A., Shafrir, Y., Hering, S., Wolschann, P., and Guy, H. R. (2008) *Channels* **2**, 210–215
31. Shafrir, Y., Durell, S. R., and Guy, H. R. (2008) *Biophys. J.* **95**, 3650–3662
32. Cuello, L. G., Jogini, V., Cortes, D. M., Pan, A. C., Gagnon, D. G., Dalmas, O., Cordero-Morales, J. F., Chakrapani, S., Roux, B., and Perozo, E. (2010) *Nature* **466**, 272–275
33. Yarotskyy, V., Gao, G., Peterson, B. Z., and Elmslie, K. S. (2009) *J. Physiol.* **587**, 551–565
34. Clayton, G. M., Altieri, S., Heginbotham, L., Unger, V. M., and Morais-Cabral, J. H. (2008) *Proc. Natl. Acad. Sci. U.S.A.* **105**, 1511–1515
35. Van Petegem, F., Clark, K. A., Chatelain, F. C., and Minor, D. L., Jr. (2004) *Nature* **429**, 671–675
36. Vitko, I., Shcheglovitov, A., Baumgart, J. P., Arias-Olguín, I. I., Murbartián, J., Arias, J. M., and Perez-Reyes, E. (2008) *PLoS One* **3**, e3560

5.2 Paper #2

FULL PAPERS

DOI: 10.1002/cmdc.200900461

Toward a Consensus Model of the hERG Potassium Channel

Anna Stary,*[a, d] Sören J. Wacker,[a] Lars Boukharta,[b] Ulrich Zachariae,[a] Yasmin Karimi-Nejad,[c] Johan Åqvist,[b] Gert Vriend,[d] and Bert L. de Groot[a]

Malfunction of hERG potassium channels, due to inherited mutations or inhibition by drugs, can cause long QT syndrome, which can lead to life-threatening arrhythmias. A three-dimensional structure of hERG is a prerequisite to understand the molecular basis of hERG malfunction. To achieve a consensus model, we carried out an extensive analysis of hERG models based on various alignments of helix S5. We analyzed seven models using a combination of conventional geometry/packing/normality validation methods as well as molecular dynamics simulations and molecular docking. A synthetic test set with the X-ray crystal structure of $K_v1.2$ with artificially shifted S5 sequences modeled into the structure served as a reference case. We docked the known hERG inhibitors (+)-cisapride, (S)-terfenadine, and MK-499 into the hERG models and simulation snapshots. None of the single analyses unambiguously identified a preferred model, but the combination of all three revealed that there is only one model that fulfils all quality criteria. This model is confirmed by a recent mutation scanning experiment (P. Ju, G. Pages, R. P. Riek, P. C. Chen, A. M. Torres, P. S. Bansal, S. Kuyucak, P. W. Kuchel, J. I. Vandenberg, *J. Biol. Chem.* 2009, *284*, 1000–1008).[1] We expect the modeled structure to be useful as a basis both for computational studies of channel function and kinetics as well as the design of experiments.

Introduction

The human ether-à-go-go-related gene (hERG) encodes the pore-forming subunits of potassium channels that conduct the rapid delayed rectifier K$^+$ current (I_{Kr}).[2, 3] I_{Kr} is activated by membrane depolarization and is a key determinant for re-polarization of the cell membrane during the cardiac action potential.[4, 5] hERG is up-regulated in various cancer cell lines, suggesting its role in the pathophysiology of cancer.[6] Mutations in the hERG gene can cause inherited long QT syndrome (LQTS), a disorder that predisposes affected individuals to life-threatening arrhythmias and sudden death.[7] Blockade of hERG can lead to acquired LQTS, a rare side effect of treatment with structurally diverse medications.[8] This potential for QT prolongation has led to severe restriction or withdrawal of several medications from the market. Intense efforts are directed at gaining a better understanding of the molecular basis of hERG channel blockade, including in vivo, in vitro, and in silico approaches (for a review, see reference [9]). Several groups have presented homology models of the hERG pore domain, providing a qualitative insight into potential ligand–channel interactions (for examples, see references [10–14]), and in some cases quantitative predictions have been provided.[15–17]

The accuracy of homology models depends critically on the sequence identity between template and target.[18, 19] The S5 helices of hERG are only distantly related to potential templates such as KcsA,[20] MthK,[21] K_vAP,[22] or $K_v1.2$,[23, 24] with sequence identities <30%. Consequently, no consensus has been established regarding the optimal alignment for S5. Differences in alignment of segment S5 have been neglected; however, S5 helices are in close contact with S6 segments, and are thus likely to influence the drug binding site. Therefore, we tested seven hERG models with different S5 alignments, five of which have been published,[11–14, 16] with a combination of state-of-the-art quality assessment methods and molecular dynamics (MD) simulations, and then analyzed the consequences of alignment errors on drug–receptor studies. To avoid potential errors and biases from quality assessment programs, a set of "control" structures, consisting of a native $K_v1.2$ crystal structure, and "artificial" models, with shifted helices, was included in our study. Generally, a quality check must be able to verify the biological relevance of a model; that is, should be able to identify improperly folded models that result from alignment

---

[a] Dr. A. Stary,$^+$ S. J. Wacker, Dr. U. Zachariae, Prof. Dr. B. L. de Groot
Computational Biomolecular Dynamics Group
Max Planck Institute for Biophysical Chemistry
Am Fassberg 11, 37077 Göttingen (Germany)
Fax: (+49) 551-2012302
E-mail: anna.stary@univie.ac.at

[b] L. Boukharta, Prof. Dr. J. Åqvist
Department of Cell and Molecular Biology
Uppsala University, Biomedical Center, Box 596, 75124 Uppsala (Sweden)

[c] Dr. Y. Karimi-Nejad
Solvay Pharmaceuticals GmbH
Hans-Böckler-Allee 20, 30173 Hannover (Germany)

[d] Dr. A. Stary,$^+$ Prof. Dr. G. Vriend
Centre for Molecular and Biomolecular Informatics
Nijmegen Centre for Molecular Life Sciences
Radboud University Nijmegen
P.O. Box 9010, 6500 GL Nijmegen (The Netherlands)

[$^+$] Current address: Department for Pharmacology and ToxicologyUniversity of Vienna, Althanstrasse 14, 1090 Vienna (Austria)

Supporting information for this article is available on the WWW under http://dx.doi.org/10.1002/cmdc.200900461.

errors. Our underlying assumption is that a reliable model (based on a correct alignment) should not fail with any assessment method, should have reasonable stability in MD simulations, and should be suitable to study drug–receptor interactions in a qualitative and possibly quantitative manner. Only one out of the seven tested hERG models fulfilled these criteria. This model was confirmed by a recent mutation scanning experiment[1] that is consistent with the alignment underlying this model.

Results

Models of the pore-forming domain

hERG models were built by using K$_v$AP as the template; they include S5 segments, the P-helix re-entrant loops, and S6 segments (Figure 1B). Alignments 2–6 were extracted from the literature,[11–14,16] and alignments for models 1 and 7 were added for completeness. During the modeling process, fourfold symmetry was applied. The S5 turret helices and voltage-sensing helices S1–S4 were omitted from the models as described previously,[13] facilitating comparison between different hERG models. Except for model 2, which was downloaded from the Schrödinger homepage (http://www.schrodinger.com/productpage/14/6/75/), the backbone of hERG models was not manually adjusted, as has been described for some models.[11,13] Farid et al.[11] describe rotational movements of backbone torsion angles for residues G648 (S6) and G572 (S5), which result in an increased pore size relative to K$_v$AP.

Figure 1 B shows the pore-forming domain of a hERG model with residues critical for drug block highlighted. Consistent with alanine scan experiments (for examples, see references [10, 25]), the side chains of residues T623, S624, Y652, and F656 of all seven hERG models are oriented toward the pore, enabling direct interactions with blockers.

In hERG, the degree of identity with templates of known structure varies for different segments. The "signature sequence" of the selectivity filter TVGYG is highly similar between potassium channels of known structure (Mlotik,[26] KcsA,[20] MthK,[21] K$_v$1.2,[23,24] and K$_v$AP[22]) and hERG, which contains a slightly modified SVGFG motif. An unambiguous alignment of S6 segments is possible due to the presence of a highly conserved glycine hinge and reasonable sequence identities of 39% between S6 segments. The outer helices formed

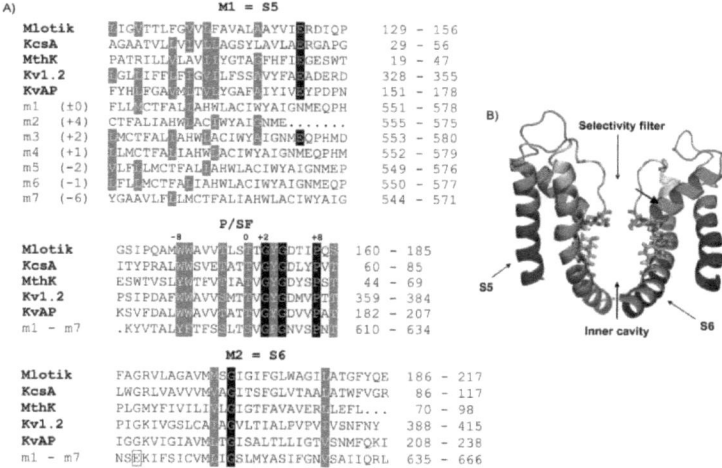

Figure 1. A) Structural alignment of the pore-forming domains of Mlotik[28] (PDB ID: 3BEH), KcsA[20] (1K4C), MthK[21] (1LNQ), K$_v$1.2[23,24] (2R9R), and K$_v$AP[22] (1ORQ) channels with hERG; m1–m7 denote the various hERG alignments with K$_v$AP. Alignments for models 2–6 were extracted from the literature (see Experimental Section). Numbers in parentheses indicate the shift of helix S5 relative to the alignment of m1 (arbitrarily taken as reference). Identical residues in all sequences are boxed in black, and similar residues are boxed in gray. Numbers above the alignment indicate residues that are part of the pore loop signature sequence present in most potassium channels;[27] E637 is boxed. B) Side view of two diagonal subunits of the pore-forming domain of hERG with key residues T623, S624, Y652, and F656 important for drug binding shown in stick representation. Helices S5 are colored blue, P-helices are colored green, the selectivity filter loops are yellow, and the S6 segments are colored red.

by S5 segments are more difficult to align due to the low sequence identity between hERG and the K$_v$AP template and generally between different potassium channels (Figure 1A). The pore domains of the potassium channel crystal structures display poor conservation at the sequence level (21–38%), but the correct alignment can be obtained by generating structural alignments. They reveal a remarkable similarity in 3D space, justifying the use of currently available crystal structures such as K$_v$AP, MthK, K$_v$1.2, or KcsA to generate hERG models. All five structures have a conserved glutamate residue at the extracellular end of S5 that interacts with backbone amide nitrogen atoms and the hydroxy group of a conserved threonine in the loop connecting the selectivity filter with S6 segments. Mutation of the corresponding residue in Shaker (E418) revealed an important role of this charged residue for normal gating.[27] Published alignments for hERG show considerable variation in S5 with relative shifts of this helix of more than one helical turn in the N- or C-terminal direction (see Figure 1A).

The alignment of model 3 suggests that E575 of hERG might stabilize the post-selectivity filter in a similar way as observed in the crystal structures. However, mutations of E575 to cysteine or positively charged lysine are well tolerated,[28] indicating a different role for E575 in hERG. All hERG models have a conserved glutamate (E637) at the top of the S6 helix, which might provide hydrogen bonds that stabilize this part of the channel (see also Stansfeld et al.[13]). This hypothesis is consistent with experimental observations that mutations of E637 to lysine lead to poor channel expression. Furthermore, mutations at this position are linked to LQTS,[29] underscoring the functional importance of residue E637.

Intra- and inter-subunit interactions in various hERG models

In all seven hERG models, hydrogen bonds between Y652 and S649 (both S6) from neighboring subunits are observed. Except for models 2 and 7, these interactions are maintained during MD simulations. Hydrogen bonds are also formed between E637, located at the extracellular end of segment S6, and the backbone of N633 and backbone and side chain of T634, located at the post-selectivity filter loop. These interactions remain stable in simulations in models 4–7, but are lost in models 1–3. Model 6 contains additional inter-domain hydrogen bonds between Y667 (S6) and T556 (S5), which are preserved in MD simulations.

Aromatic–aromatic interactions

In the hERG pore domain, 15–19 aromatic residues per subunit are present, depending on the alignment used. Five to eight of these are present in S5 segments. Alignment differences lead to different interaction patterns between aromatic side chains in the models (Table 1 and Figure 2). Model 1 has the greatest number of favorable aromatic–aromatic interactions, with four pairs between residues of S5, the P-segment, and S6 from the same subunit. Additionally, a cluster of three aromatic side chains between S5, the P-helix, and the selectivity filter from the neighboring subunit is present. In model 6, two clusters

Table 1. Interactions of S5 aromatic residues with other aromatic residues.

Model	F551	F557	H562	W563	W568
m1	–	Y667 (S6)	F619 (P)	Y611 (P)	F617 (P) F627 (SF) inter[a]
m2	–	–	–	F619 (P)	F640 (S6)
m3	Nn[b]	–	–	Y611 (P)	–
m4	Nn[b]	–	–	–	–
m5	–	F619 (P)	–	F617 (P)	–
m6	–	Y652 (S6) F619 (P)	Y611 (P)	–	–
m7	Y667 (S6)	–	F640 (S6)	F617 (P)	–

[a] Inter-subunit interaction. [b] Not in model.

with three aromatic side chains and one paired interaction between S5 and the P-segment exist. In model 7, three pairs of aromatic–aromatic interactions are identified, while models 2 and 5 possess only two aromatic pairs, between S5 residues and neighboring segments. Models 3 and 4 contain only one aromatic pair between S5 residues and other segments.

Model evaluation—quality assessment programs

To identify the correct alignment, various quality assessment methods have been applied. Because most of the methods have been developed for globular proteins and use statistical potentials from known structures, we validated the suitability of the selected quality assessment methods for membrane-spanning potassium channels. The crystal structure of K$_v$1.2 was used because it has higher resolution (2.4 Å) than that of K$_v$AP (3.2 Å). K$_v$1.2 scores well with all methods tested; no low scores are reported. Scores for K$_v$AP are also acceptable, with the exception of Verify3D and Procheck, with which low values are observed. To assess the discriminative power of these programs to distinguish correctly aligned from misaligned models, a synthetic test set consisting of K$_v$1.2 models with shifted S5 segments (one helical turn in steps of one amino acid in both directions, m−1 denotes a shift toward the C terminus) was built, and the quality screened. Results are summarized in Table 2A. Only four out of eight methods (WHAT_CHECK Packing 2, Prosall, ProQres LG, and DFIRE) ranked the crystal structure highest. Most shifts in S5 do not noticeably influence the results obtained with ProQres, Procheck, or ModFOLD. The strongest variations were obtained with Verify3D, with which the best model (alignment shifted two residues toward the N terminus) scored ~23% higher than the worst model (alignment shifted three residues toward the C terminus), but the crystal structure was ranked only third best (Table 2A).

Table 2B summarizes the results obtained with Verify3D, Procheck, WHAT_CHECK, Prosall, ProQres, DFIRE, and ModFOLD

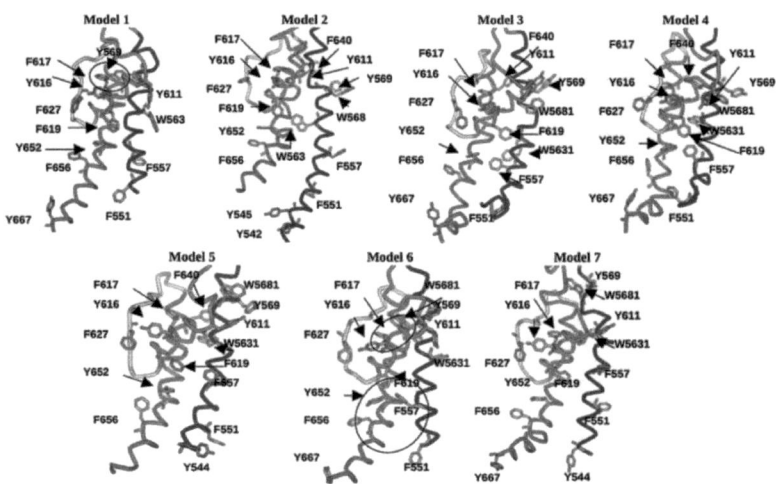

Figure 2. Details of aromatic–aromatic interactions in the various hERG models 1–7: Shown in each case is a side view of averaged hERG coordinates after 10 ns MD simulations, with aromatic side chains (Phe, Tyr, Trp) shown in stick representation. S5 segments are colored blue, P-helices are shown in green, the selectivity filter is yellow, and S6 segments are colored red. For model 1, favorable aromatic–aromatic interactions between segment S5, the P-helix, and S6, stabilizing the selectivity filter are circled. For model 6, only favorable aromatic–aromatic interactions between S5 and other segments are observed. Stabilizing interactions in the selectivity filter region and the pore region are circled.

Table 2. Static quality assessment of crystal structures plus A) synthetic test set and B) hERG models.[a]

A)

Model	Prosa2003		Verify3D		ProQres LG score		ProQres MaxSub		WHAT_CHECK Packing 2		DFIRE/res.		Procheck ϕ/ψ		ModFOLD	
KvAP	−5.57	–	68.88	–	5.431	–	0.201	–	−0.447	–	−100.55	–	83.5	–	0.80	–
Kv1.2	−5.55	6	76.08	3	6.880	1	0.636	2	−0.172	1	−109.59	1	92.8	9	0.89	6
tm+1	−3.67	5	71.76	5	5.977	5	0.617	3	−0.736	8	−103.61	4	97.6	1	0.93	2
tm+2	−3.71	2	86.26	1	5.740	6	0.606	4	−0.724	7	−102.27	7	96.1	5	0.92	3
tm+3	−3.41	8	66.41	6	6.648	2	0.573	7	−0.529	3	−103.20	5	97	2	0.91	5
tm+4	−3.07	9	55.98	9	5.021	8	0.472	9	−0.699	6	−102.27	8	95.5	6	0.86	7
tm−1	−4.1	1	75.83	4	6.207	4	0.600	5	−0.280	2	−105.69	2	94.9	7	0.93	1
tm−2	−3.7	4	77.35	2	6.403	3	0.679	1	−0.652	4	−104.94	3	96.1	4	0.92	4
tm−3	−3.71	3	62.85	8	5.655	7	0.583	6	−0.696	5	−102.89	6	94.9	8	0.82	8
tm−4	−3.42	7	66.41	6	4.723	9	0.494	8	−0.858	9	−100.79	9	97	3	0.82	9

B)

Model	Prosa2003		Verify3D		ProQres LG score		ProQres MaxSub		WHAT_CHECK Packing 2		DFIRE/res.		Procheck ϕ/ψ		ModFOLD	
m1	−4.01	2	51.46	5	4.985	1	**0.145**	6	−1.613	4	−103.46	3	93.2	2	0.65	3
m2[10]	−2.09	5	39.22	7	**1.622**	7	**−0.081**	7	−0.606	1	**−88.69**	7	**89.2**	7	**0.32**	7
m3[11]	−2.41	4	64.38	3	4.266	3	0.380	2	−2.078	7	−99.67	6	92.8	3	**0.38**	6
m4[15]	−2.01	6	69.77	2	3.922	4	0.365	3	−2.057	6	−99.69	5	93.4	1	**0.38**	5
m5[12]	−1.99	7	50.96	6	3.520	5	0.276	5	−2.016	5	−100.16	4	92.2	5	0.41	4
m6[13]	−3.58	3	79.2	1	4.377	2	0.408	1	−0.855	2	−105.92	2	91.2	6	0.68	1
m7	−4.75	1	57.84	4	3.418	6	0.339	4	−1.530	3	−115.57	1	92.4	4	0.68	1

[a] Models are ranked 1–7, with 1 denoting the best-ranked model in each category; low values are shown in boldface.

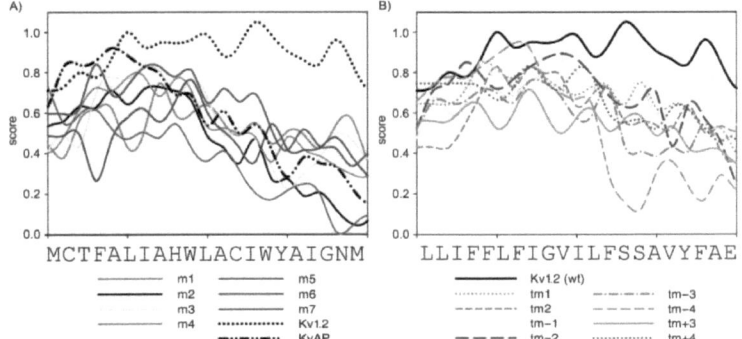

Figure 3. Local model quality assessment: A) Local quality of segment S5 of all seven hERG models, as well as crystal structures K$_v$AP and K$_v$1.2, as assessed with the method developed by Fasnacht et al.[30] The amino acid sequence of helix S5 is shown below the plot. Higher scores indicate higher quality. B) Local quality of K$_v$1.2 and synthetic test models.

for the hERG models. Models are ranked from 1 to 7, with 1 denoting the best-ranked model in each category. Low values are shown in bold. Static checks show reasonable scores for model 1, with the exception of Verify3D and ProQres MaxSub score, which are lower than expected for good structures. Scores for model 2 are very low, indicating severe problems. Models 3–5 have similar quality, with problematic packing values and very low scores with ModFOLD. Scores for model 6 are favorable, and none of the methods suggest structural problems. Scores for model 7 are acceptable except for Verify3D, which reports a low value (see Table 2B). The ranking of hERG models is less clear, and lower scores are obtained for all models relative to the crystal structures. Model 6 scores slightly better than models 3 and 4, model 1 has intermediate quality, and models 5 and especially 2 score lowest.

The local quality of S5 segments was evaluated with a method developed by Fasnacht et al.[30] The results are summarized in Figure 3. K$_v$1.2 shows the highest quality throughout the whole segment. This method shows a clear distinction between different alignments in our synthetic test set, with the largest differences observed for the second half of S5. The results for the hERG models are less straightforward to interpret, yet a similar quality trend emerges. Models 1 and 6 are again among the best, models 3, 4, and 5 show intermediate qualities, and models 2 and 7 score lowest.

Molecular dynamics simulations

hERG models were examined using MD simulations with the protein models embedded in a POPC lipid bilayer, and each simulation was repeated twice with different initial velocities. The root mean square deviation (RMSD) of a protein from its starting coordinates as a function of time is routinely used as a measure of its structural stability. Figure 4A shows the RMSD values of the backbone atoms of the hERG models. The stability was compared with K$_v$AP and K$_v$1.2 (Figure 4B), which display RMSD values in the range of 0.16–0.2 nm. hERG models 1, 3, 5, and 6 have only slightly higher RMSD values in the range of 0.2–0.25 nm. Models 2, 4, and 7 are less stable, with RMSD values in the range of ~0.35 nm after 10 ns. Furthermore, the RMSD curves of these models are still rising, indicating that these systems have not yet found local minima. The results for the extended simulations of model 1 (60 ns) and model 6 (100 ns) are shown in Figure 4C.

Table 1 lists aromatic–aromatic interactions in the various hERG models prior to MD simulations. Important changes are observed during MD simulations; these variations are of particular interest, as they may directly influence drug blocking. Figure 5 summarizes the distance measurements between F656 from various subunits as a function of time. The distance is defined as the distance between the geometric centers of the phenyl rings.

Models 1 and 5 show very short distances between adjacent phenyl rings. Models 3, 4, and 7 show values in the range of 1–1.5 nm, and model 2, which has been adjusted manually to increase the pore size (for details, see reference [11]) and model 6 (no modification) display larger inner cavities, with distances between phenyl rings in the range of 2–2.5 nm. In models 3 and 5, the inner cavity "collapses" during MD simulations; F656 residues from all four subunits cluster tightly.

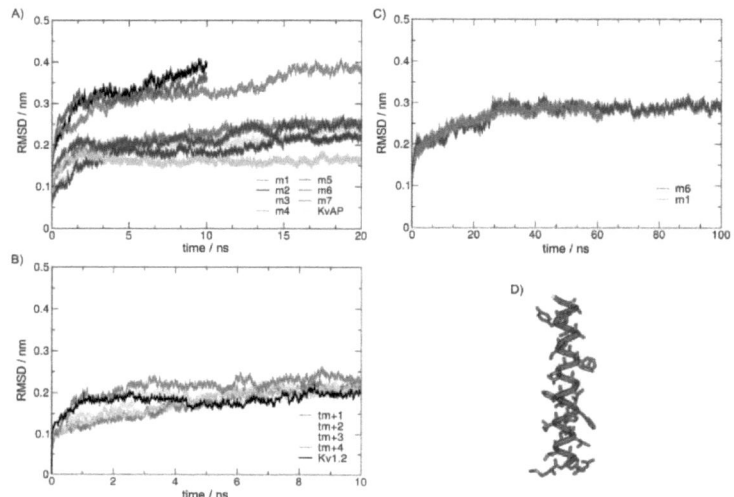

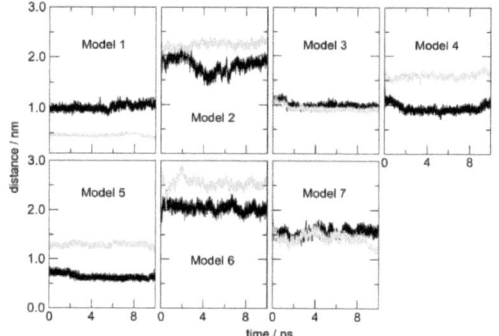

Figure 4. RMSD values of the various hERG models, K$_v$AP, and K$_v$1.2 plus synthetic test set: A) Results from MD simulations with hERG models embedded in a POPC lipid bilayer under physiological ion concentrations are shown. All simulations were repeated twice, as described in the Experimental Section. The RMSD plots for all backbone atoms of hERG models m1–m7 are shown. B) The stability of K$_v$1.2 and different K$_v$1.2 test models during MD simulations. C) Simulations for models 1 and 6, which performed best in our static assessments, were simulated for 60 and 100 ns, respectively. D) Ribbon and sticks representation of helix S5 from the K$_v$1.2 crystal structure.

Figure 5. Influence of the conformation of F656 on hERG pore size: The distances between the geometric centers of F656 phenyl rings of adjacent subunits are shown as a function of time. Models 1, 3, 4, and 5 show very narrow pores, with F656 residues from several subunits interacting directly with each other, thereby "collapsing" the pore. Larger distances between adjacent F656 rings are observed for models 2 and 6, while model 7 displays an intermediate pore size.

Experimental validation of hERG models

Recently, Ju et al.[1] studied the structure and function of helix S5 in detail using a combination of NMR and mutagenesis studies. These data were not included in the model building and evaluation process and can therefore be used to cross-validate the results of our study. Figure 6 shows a side view of one domain of each hERG model, with residues perturbing inactivation[1] shown as spheres. According to Ju et al.,[1] these residues should point toward the inner S6 helix. Using the alignment of model 1, only F551 and the side chain of W568 are orientated toward the pore. In model 2, the situation is even

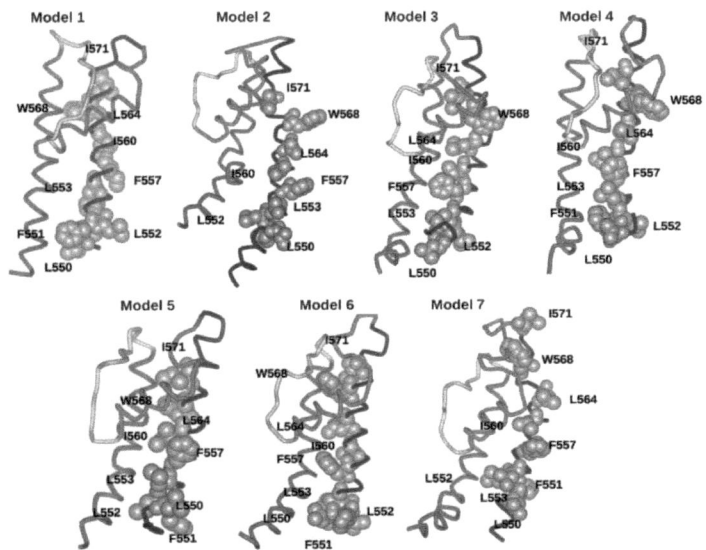

Figure 6. Agreement of the various hERG models with the results of a recent mutagenesis study: Shown are ribbon representations of one subunit of each hERG model, with residues perturbing the inactivation shown as spheres; these residues are thought to be predominantly pore-facing.[1]

worse; except for L552 all inactivation-perturbing residues are orientated toward the voltage-sensing domain. In model 3, only residues L551 and W568 face the opposite side of the pore. In model 4, two of the inactivation-perturbing residues, L552 and W568, are orientated toward the voltage-sensing domain. In model 5, residues L550 and F551 cannot interact with the inner pore helix. Model 6 fits the experimental data of Ju et al.[1] best, as only the side chain of L552 does not face toward helix S6. The alignment of model 7 is shifted extensively, and residues W568 and I571 are not in the transmembrane core, but in the extracellular loop region. Additionally, the side chains of F551 and L564 are orientated toward the voltage-sensing domain.

F557 on S5 was found by Ju et al.[1] to exert an especially pronounced facilitating effect on the inactivation of hERG. Mutation of this residue to alanine decreases the energy barrier to inactivation by ~1.5 kcal mol^{-1}. In our model 6, the side chain of F557 is situated next to that of Y652 of S6, and both aryl rings undergo a direct π–π stacking interaction in the model which remains stable in MD simulations. It is conceivable that F557 influences the rotameric state of the Y652 side chain in hERG. In turn, Y652 stacks in a parallel fashion onto F656 and thus stabilizes its conformation. The aromatic residues on S6 have been shown to be strongly involved in inactivation and/or drug binding.[31,32] In further support of model 6, residues within the helical part of S5 (F551, L559, and W563) experimentally found to facilitate activation gating are orientated toward the voltage-sensing domains. A strong interaction with the voltage-sensing domain thus appears highly plausible, as was suggested by Ju et al.[1]

Interaction of different hERG models with high-affinity blockers

A large amount of experimental data on various drugs is available for hERG. Using these data to distinguish between various S5 alignments is not straightforward, as mutational studies focus on the inner S6 segments and the bottom of the selectivity filter, where drug binding occurs. However, the alignment of both SF and S6 segments are identical for all seven hERG models. An indirect evaluation based on proposed drug binding modes might be possible. For this purpose, we docked the high-affinity blocker (+)-cisapride into the cavities of all seven hERG models. Cisapride was selected because the importance of the positioning of the aromatic residues in S6 has been demonstrated experimentally (for examples, see references [31,33]). Additionally, (S)-terfenadine and MK-499 were docked into the two best-ranked models, model 1 and model 6. FlexX and GOLD, with standard parameters, were used to analyze hERG–drug interactions. Both programs require the definition of residues that line the binding cavity. Therefore, residues Y652, F656, T623, and S624 from all four subunits were chosen as starting points for docking. Observations in the docking of cisapride are described in detail for all seven models (Figure 7). Additionally, details of the docking results for (S)-terfenadine and MK-499 are presented for model 6, but not for model 1, because docking was not successful in the latter case. Minimized averaged structures after 10 ns MD simulations as well as structures prior to MD simulations served as starting conformations for docking, and only the 20 top-scoring poses of each model were analyzed. For models 1, 3, and 5, no reasonable docking poses within the cavity could be obtained using minimized average structures after MD simulations. Therefore, additional minimized snapshots of the trajectory were extracted for docking. However, most snapshots were unsuitable for docking, with the exception of the starting coordinates, which provided large enough cavities for drug docking (data not shown).

The docking poses for the starting structure of model 2 are similar as described in Farid et al.[11] In agreement with this study, cation–π interactions were not predicted. Docking poses obtained from the average structure after 10 ns MD simulation changed due to considerable deviations from the starting structure (Figure 4), and a hydrogen bond between the methoxy group of the benzamidine ring and the hydroxy group of S649 (S6) was predicted. Aromatic interactions of Y652 and F656 to three subunits are observed; however, instead of interactions to three F656 and two Y652 residues from different subunits, interactions to two Y652 and three F656 residues are predicted for the final MD structure.

Docking scores obtained for model 4 are considerably lower than the results for model 2. Again, due to large deviations

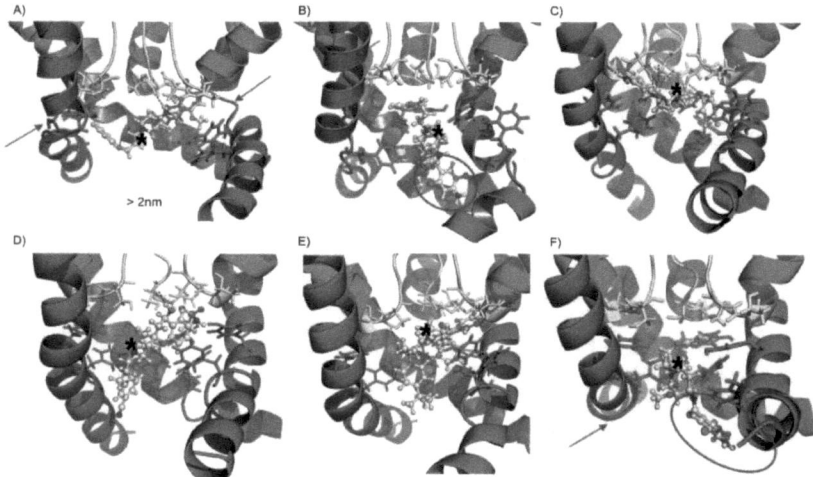

Figure 7. Interactions of the various hERG models with the high-affinity blocker cisapride. Snapshots of the best-ranked docking poses for models 2, 4, 6, and 7 are shown. Helices S6 (red), P-helix (green), and the selectivity filter (yellow) of two diagonal subunits are shown in ribbon representation, and residues T623, S624, Y652, and F656 are shown as sticks and are colored according to the segments to which they belong. The location of the positively charged nitrogen atom is marked (*). A) Red arrows indicate the distorted geometry of the α helix in this region, which is most likely a consequence of the interventions at G648. Hydrogen bonds are shown as black dotted lines. B) Orientation of cisapride in the averaged coordinates (10 ns) of model 4. Helices S6 are kinked, influencing the shape of the binding cavity. The benzamidine ring of cisapride (circled in red) does not interact with any experimentally determined residues, but is oriented toward the intracellular cavity. C) Cisapride interactions with averaged coordinates of model 6 after 10 ns MD simulation. Hydrogen bonds are shown as black dotted lines. D) Interactions of MK-499 with the averaged coordinates of model 6 after 10 ns. The hydrogen bond to S624 is shown as a black dotted line. E) Terfenadine hERG interactions for model 6, after 10 ns MD simulation, are shown. Hydrogen bonds to the selectivity filter residues are shown as black dotted lines. F) Best docking pose obtained for the averaged coordinates of model 7 after 10 ns MD simulation. Helices S6 are considerably bent, influencing the shape of the cavity. Similar to model 4, the drug is orientated perpendicular with respect to the benzamidine ring of cisapride oriented toward the intracellular side (circled).

from the starting structure, (+)-cisapride–hERG interactions changed considerably from the starting structure to the averaged structure after 10 ns MD simulation. Only results for the averaged structure are described in detail in the following section, as the cavities in all starting structures, except for model 2, are very similar. No hydrogen bonds between (+)-cisapride and any polar residues in the cavity are predicted. Interactions to five aromatic side chains are predicted: three to Y652 and two to F656. The benzamidine ring interacts via parallel displaced π–π stacking from neighboring Y652 residues, while hydrophobic contacts are predicted to F656 residues.

Docking into starting and end coordinates of model 6 yielded reasonable scores. Simultaneous interactions between six aromatic side chains and (+)-cisapride are observed. The fluorophenoxy ring is flanked by two Y652 rings and one F656 ring which interact via T-shaped π–π stacking. Furthermore, a hydrogen bond is predicted to the oxygen atom of one of the serines (S624) from the base of the selectivity filter. Hydrogen bonds are also predicted between methoxy groups and two additional S624 residues and to the backbone carbonyl oxygen atom of one T623 residue. The benzamidine ring interacts with Y652 via parallel displaced π–π stacking and F656 from the same subunit via T-stacking. There are more favorable interactions between (+)-cisapride and model 6 than to any other model.

In contrast to cisapride, only two hydrogen bonds between one hydroxy group of (S)-terfenadine and the side chains of S624 and (a weaker hydrogen bond) to T623 from the same subunit are predicted. This result is consistent with a recent alanine scan reported by Kamiya et al.[25] Additionally, π–π interactions to three Y652 residues and one F656 are predicted. This finding is not completely consistent with the study reported by Imai et al.,[33] in which interactions to diagonal but not adjacent Y652 residues are suggested.

Interactions between model 6 and MK-499 are illustrated in Figure 7. In agreement with a study by Mitcheson et al.,[10] a hydrogen bond between S624 and the hydroxy group of MK-499 is observed. However, the distance is much smaller (1.8 Å) as proposed by a recent docking study.[34] Similar to the findings of Farid et al.,[11] interactions between MK-499 and four aromatic side chains are predicted. No direct contacts to G648 are observed (see Figure 7).

The averaged structure after 10 ns MD simulation provided a suitable starting conformation to probe (+)-cisapride interactions with model 7. Similar to model 4, docking scores are significantly lower than for models 2 and 6. The binding site in model 7 differs from other models, because bending of the S6 segments in the region between Y652 and F656 occurs, and the cavity is generally smaller than in models 2 and 6 (see Figures 2 and 5). Again, results for the averaged model after 10 ns are quite different from the starting coordinates, as the model is not stable in MD simulations. No hydrogen bonds between selectivity filter residues (T623, S624) and (+)-cisapride were observed. Interactions with six aromatic side chains from three different subunits are observed. Aromatic interactions with the fluorophenoxy ring are predicted via parallel stacking between two Y652 residues from neighboring subunits and sandwich stacking occurs to an F656 residue. Surprisingly, no aromatic contacts to the benzamidine ring are present, and the basic nitrogen atom faces toward the intracellular cavity. Taken together, models with collapsed or very narrow inner cavities (models 1, 3, and 5) are not suitable for drug docking. The most favorable docking poses were obtained for model 6, followed by model 2. Models 4 and 7 have fewer favorable contacts than the other models.

Discussion

In the present study we investigated the quality and stability of seven different hERG models obtained by homology modeling. Due to the low sequence identity in S5 helices, no consensus about the alignment of this segment has been achieved (Figure 1 A). It was therefore a main goal of this study to identify the correct alignment of helix 5.

The importance of aromatic–aromatic interactions in ion channels has been noted previously.[20,35] In KcsA, residues W67 and W66 form an aromatic cuff, stabilizing the pore loop.[20] This motif,[36] which is part of the pore loop signature sequence, is conserved in hERG. The exception is position +8, which is replaced by a tyrosine residue (see Figure 1 A).

A common observation in membrane proteins is the preferential location of aromatic residues, especially Trp and Tyr at the interface between membrane and solvent.[37–41] In the hERG models, aromatic residues are not only at the membrane interface in most models (see Supporting Information figure 1), but there are additional aromatic residues distributed over the protein, making this feature difficult to interpret. Evenly distributed aromatic residues have been observed in the closed conformation of Kirbac1.1. However, a shift toward extra- and intracellular regions has been suggested for the open conformation.[40]

There are significant differences between aromatic–aromatic interactions in different hERG models (Table 1 and Figure 2). Interactions between aromatic residues are energetically favorable.[41] One could therefore expect a contribution to the stability of the protein. However, we found limited correlation between the aromatic clusters and the structural stability in MD simulations. Models 1 and 6 are indeed among the most stable, but model 7 is among the least stable (Figure 4). Upon close inspection, an aromatic mismatch between model 7 and the lipid bilayer becomes apparent. With lipid bilayers such as POPC, it is not possible to satisfy the location of most Tyr and Trp residues in the membrane head group region. Several Tyr groups extend into the bulk water. This might well explain the high RMSD values, despite favorable aromatic–aromatic interactions in the protein structure per se.

The assessment methods used show a clear trend among the studied models. Model 2 scores very low with all but one of the tested methods, whereas model 6 is the only model that does not show low values for any of the applied checks. Models 3–5 show intermediate quality, and scores for model 7 are better than average with the exception of Verify3D, which indicates low quality (Table 2 A). To evaluate the reliability of the methods used to discriminate between slightly different models, we introduced a synthetic $K_v1.2$ test set, with shifted helices. We found limited discriminative power for this decoy set (Table 2B). Part of the reason for this is that none of the methods used was developed specifically for membrane proteins. Programs such as Verify3D take the local residue environment into account. Therefore, limited use of this method might be expected, because the environment for membrane proteins differs significantly. The membrane environment is predominantly hydrophobic with little possibility for hydrogen bonding and electrostatic interactions. This is also reflected in the differences in amino acid composition[42] with different secondary-structure propensities between membrane environments and aqueous solution.[43] Indeed, results obtained by Verify3D are not able to discriminate between $K_v1.2$ and the decoy set. We find that DFIRE and the local assessment method from Fasnacht et al.[30] are able to discriminate between different alignments (Table 2 and Figure 3), whereas methods such as Procheck (ϕ/ψ angles) are not suited for this task.

Law et al.[44] found a good correlation between quality as assessed with static structure assessment and stability in MD simulations. Whether MD simulations are suitable to distinguish between correctly and incorrectly folded models, for example, via misalignment of certain segments is still an open question. If helices are incorrectly packed one might expect to see a greater degree of drift from the starting coordinates reflected in large RMSD values.[45-47] Simulations have also proven helpful to distinguish between different Kir6.2 alignments,[44] and we have previously shown that a distinction between clockwise and counterclockwise orientation in an L-type calcium channel model is possible.[48] On the other hand, Law et al.[44] report that it is not possible to distinguish between different subunit orientations in Twk-5 channels. Clearly this issue is inconclusive. In hERG models, large drifts (i.e., repacking of helices) were observed only for three out of the seven models (Figure 4A). The simulations suggest that one can identify models that contain serious problems, such as model 2, which scored poorly in most of the static assessments and also displays a rather high RMSD value. However, models 1 and 6, which are ranked among the best models, are difficult to distinguish using MD simulations. Simulations for these two models were extended to 60 (m1) and 100 ns (m6), with still no clear distinction between them (Figure 4C). This suggests that although MD can identify poor model quality, low RMSD values do not automatically identify the correct fold.

Different RMSD values reflect packing differences between S5 and S6/P-helix. The exception is model 2, which has a somewhat longer sequence (S5 intracellular), but it is unlikely the sole reason for the higher RMSD values observed for this model. At the end of the 10 ns simulation, no equilibrium is reached.

Significant differences in stability could not be detected in our $K_v1.2$ decoy test sets, in which we mimicked incorrect packing between the outer M1 (corresponding to S5) and inner M2 (corresponding to S6) helices, by introducing artificial shifts in the outer helices (Figure 4B). This might be at least partially explained by the fact that M1 is rather symmetric, and changes in helix–helix packing upon shifts are relatively minor (Figure 4C).

In three out of the seven hERG models (m1, m3, and m5) a drastic decrease in cavity size was observed due to a collapse of all four aromatic F656 residues. This effect has been described previously for model 3.[12] The authors reported the unsuitability of the model for docking studies after 3.5 ns simulation, which is in agreement with our analysis. It is reasonable to assume that the orientation of the pore residues varies with the channel state (i.e., inactivated—activated—deactivated; for examples, see references [31–33]), and we cannot exclude that this "collapse" represents such a state. However, repeated collapses on a very short time scale (several nanoseconds) are at least suspicious, as although inactivation in hERG is fast, it occurs on the millisecond time scale. Additionally, MD simulations of a high-resolution crystal structure of the NaK cation channel in the open conformation, which also contains four pore-facing Phe residues, did not lead to similar narrow cavities (our unpublished observations). Our study suggests that there is a correlation between pore collapse (see Figure 5) and (mis)alignment.

Experimental validation of the S5 segments comes from a recent alanine scan by Ju et al.[1] These data agree best with the alignment of model 6 and thus confirm this model as likely open conformation hERG structure (Figure 6). This alanine scan is least compatible with model 2, which was constantly ranked lowest with various programs.

Ju et al.[1] show that mutations on the S5 helix are capable of strongly interfering with the energetics of inactivation and activation gating in hERG. This underscores the importance of modeling this section of hERG in addition to the cavity-forming helices S6, P, and the selectivity filter, in order to study the molecular basis of hERG kinetics and its possible impact on drug binding, as inactivation plays an important role in high-affinity drug interactions.

Most residues on S5 experimentally shown to impact inactivation energetics were found to face S6 in our model. The effect of the mutant F557A is especially intriguing, because it decreases the barrier toward inactivation substantially. The phenyl ring of F557 directly faces the side chain of Y652 on S6 in our model. Recent experimental work by Klement et al.,[32] in which the effect of the hERG-like mutation I470Y on the inactivation properties of Shaker (I470 is homologous to Y652 in hERG) was studied, shows that the mutation effects a considerable acceleration of the C-type component of inactivation. Klement et al.[32] suggest that it is the rotameric state of the tyrosine side chain (cavity-pointing vs. cavity-lining) which is most crucially involved in the induction of this inactivated state. The close stacking interaction between Y652 and F557 observed in our model could serve as a basis to explain the strong effect of this S5 residue on inactivation, as mutation of F557 most probably influences the conformational state of Y652 in this environment.

Very recently Lees-Miller et al.[49] reported interactions of segment S5 with the pore helix of hERG. The alignment in their paper corresponds to the alignment of model 1 described herein. This model scores second best in our overall analysis; however, it is in contradiction with experiments reported by Ju et al.[1] Although the overall structure of model 1 was stable in

MD simulations, the pore-facing F656 residues were found to collapse during MD simulations. In our model 1, the proposed hydrogen bonding network between H562 from S5 and T618 and S621 from the P-helix was not observed. We therefore rebuilt model 1 using the newest Modeller version (9v6), but again, in none of the 100 generated structures were hydrogen bonds between H562 and T618 and S621 present. To address the question if such a putative hydrogen bonding network would influence the stability of the inner pore, that is, prevent collapse of the F656 rings, we performed two independent MD simulations of 20 ns including distance restraints (force constants 10 and 100 kJ mol^{-1} nm^{-2}) to enforce similar hydrogen bonds as suggested by Lees-Miller et al.[49] In both simulations, only T618 was able to form hydrogen bonds with H562; however, S621 was found to be too far away (~6.4 Å) from the histidine side chain. Furthermore, no influence on the stability of the inner cavity was observed. It is not clear if model 6, as proposed by our work, can explain the mutational data on H562, because in this model residue H562 is orientated toward the voltage-sensing domain, which was not modeled.

Additionally, we performed a limited docking study on the well-studied high-affinity blockers (+)-cisapride, (S)-terfenadine, and MK-499.[10,25] The outcome of our limited drug–receptor evaluation is, with the exception of model 2, in good agreement with results obtained by static assessment methods and MD simulations. Docking into model 6, which scored best with most assessment methods, yielded docking results that are in good agreement with alanine scan experiments.[10,25] Results for model 6 are also partly in agreement with recent chimera studies by Imai et al.[33] and Myokai et al.,[50] in which the nature of aromatic interactions for cisapride and terfenadine were studied. In agreement with Kamiya et al.,[25] hydrogen bonds to the selectivity filter residues are predicted.

No hydrogen bonds between the phenol side chain of Y652 and any of the three studied high-affinity drugs were predicted, which agrees with the study by Fernandez et al.[51] However, docking into static models cannot explain the importance of V625 or G648, which influence the affinity of MK-499.[10] It might be possible that these residues interact with blockers via an indirect mechanism.

Cisapride was docked into all seven models, but did not fit into the collapsed pores of models 1, 3, and 5. In model 1, which scored second best in our quality assessment, two additional drugs (terfenadine and MK-499) were docked with two different programs. However, none of the tested drugs could be accommodated in the narrow cavity. This is surprising because there is good evidence that certain drugs such as MK-499 might be trapped in the closed channel pore and thus could possibly fit into a narrow cavity. Furthermore, docking studies, with blockers in the closed hERG channels have been published (for examples, see references [10,13,15,34]). This prompted us to compare the cavities of model 1 with a closed channel cavity. Surprisingly, the volume of the collapsed pore is smaller than the cavity of a closed channel pore (our unpublished observations). Therefore, we suggest a possible relationship between misalignment and pore collapse in MD simulations. In agreement with this hypothesis, the size of the inner cavity for model 6 does not change significantly over time, and no collapse of the pore (up to 100 ns) emerges.

Surprisingly, model 2 performed quite well in docking analysis, despite the poor values obtained in quality assessments. It is likely that the results for model 2 are influenced by the adjustments of glycine residues in S5 and S6, leading to an artificially large inner cavity.[11] Reasonable binding modes could be obtained for the best and the lowest scoring model, highlighting the limits of static docking methods to validate models. One major disadvantage of such methods is the limited possibility to address receptor dynamics, which will be an essential step toward understanding the promiscuity of hERG. The model of the hERG structure presented in this study provides a basis for addressing the relation between ligand affinity and hERG conformational dynamics, using methods that take the receptor dynamics explicitly into account. In particular, the model can help to study aspects of channel kinetics such as mechanisms inducing entry into inactivation and activation gating in greater detail. Dependence of high-affinity drug binding on conformational changes related to inactivation gating has been described before,[50,52] and some members of the relatively newly discovered class of channel activators appear to work through inhibition of channel C-type inactivation,[53] whereas others seem to prevent deactivation.[54a] The elucidation of these conformational changes also requires knowledge of the electrostatic properties of the entire pore region. A complete and reliable molecular model of the pore-forming part of hERG is a prerequisite to understand these mechanisms and their possible influence on the characteristic promiscuity of drug binding.

Limitations

The major aim of this study was the identification of the correct alignment for segment S5 in hERG. We did not focus on detailed refinement of the models, and although model 6 fits the experimental data for S5 well, we cannot rule out that further refinement might be necessary, for example, to study inactivation. Furthermore, structurally important segments such as the S5 turret helices and the voltage-sensing domains are missing. Moreover, the structural quality of the models presented herein is limited by the resolution of the K$_v$AP template (3.2 Å). Nevertheless, we view the structural model presented here as an important step, as it represents the most plausible model of the hERG inner pore structure based on currently available data, and provides a necessary prerequisite to study the determinants of ligand binding to the hERG inner cavity.

Conclusions

It is critically important to use a combination of methods to assess the quality of homology models.[44] Careful model evaluation is crucial, especially when target and template are distantly related (i.e. below the 30% identity threshold). With a combination of static assessment programs, MD simulations, and experimental validation, we identified the most likely alignment for hERG out of seven suggested possibilities. Our study clearly shows that a careful evaluation of model quality is able to distinguish between different alignments. Furthermore, we show that alignment errors, even in segments not directly involved in drug interactions, can severely influence the shape and size of the binding site. Using this combined approach, we propose a consensus model of the hERG potassium channel structure that can be used as a basis for structure-based ligand affinity predictions, to study structure–function relationships, and to inspire future experiments.

Experimental Section

Model building

We used Modeller 7v7[55] to generate homology models of the open conformation of the human ERG1 (accession number: Q12809) channel using the K$_v$AP crystal structure (PDB ID: 1ORQ) and a refined model thereof[56] as templates. K$_v$AP was preferred over K$_v$1.2 because the latter channel contains a PXP motif in the inner S6 segments, which might change the shape of the pore.[51] Furthermore, the sequence of K$_v$AP is more similar to that of the hERG pore domain. Alignments 2–6 were extracted from published sources,[11–14,16] and alignments for models 1 and 7 were added for completeness. Fourfold symmetry was imposed for modeling the tetrameric structures of the pore-forming domains including S5 segments, P-helices, re-entrant loops, and S6 segments. Models do not include the S5-P linkers and the voltage-sensing domains. Additionally, a synthetic test set containing the K$_v$1.2/2.1 chimera pore domain (PDB ID: 2R9R) plus eight models with shifted S5 segments were built.

Static quality assessment

Verify3D,[57] Procheck ϕ/ψ angle check,[58] WHAT_CHECK Packing 2,[59] Prosa2003,[60] ProQres,[61] DFIRE,[62] ModFOLD,[63] and a local quality assessment method developed by Fasnacht et al.,[30] were used to assess the quality of various models.

Brief description of quality assessment methods

Verify3D[57] is a knowledge-based method that uses statistical potentials from real proteins and assesses how well a sequence fits its 3D structure by taking into account the residue environment combining secondary structure, solvent accessibility, and polarity. Procheck[58] assesses the stereochemical parameters of a protein. The Ramachandran plot shows the ϕ/ψ torsion angles for all residues in the structure, defining different regions populated to very unusual or "forbidden" values. WHAT_CHECK Packing[59] uses "fixed fragments" in a protein structure and checks the occurrence of all possible atom types in all possible positions around these fragments. Frequently occurring configurations are considered preferred. A summary score for each residue is calculated. Prosa2003[60] uses distance- and surface-dependent statistical potentials for C$^\alpha$ atoms of all residues in the model. ProQres[61] is a structure-based method that analyzes atom–atom contacts, residue–residue contacts, solvent-accessible surfaces, and secondary structure. The DFIRE score is a statistical potential summed over all pairs of non-hydrogen atoms. As reference state, DFIRE[62] uses a distance-scaled finite ideal gas. The ModFOLD method,[63] which is available as web sever, combines data from ModSSEA,[30] MODCHECK,[64] and ProQ[65] using a neural network to predict the accuracy of a model. The method developed by Fasnacht et al.[30] uses a combination of different statistical potentials (DFIRE, contact and torsion potentials), and structural features making use of programs such as DSSP,[66] psipred,[67] and Verify3D, using a support vector machine to assess the local quality of a model.

Molecular dynamics simulations

MD simulations were performed with Gromacs v. 3.3.[68] All hERG models, as well as crystal structures of K$_v$AP and K$_v$1.2 plus synthetic models, were embedded in an equilibrated simulation box of 241 palmitoyloleoyl phosphatidylcholine (POPC) lipids. The channels were inserted into the membrane as described previously.[69] K$^+$ ions were placed in the channel at K$^+$ sites S0, S2, and S4, with waters placed at S1 and S3 of the selectivity filter.[70] Cl$^-$ ions were added randomly within the solvent to neutralize the system. Identical simulations with an ionic strength of 150 mM were also carried out. Lipid parameters were taken from Berger et al.,[71] and the OPLS-all-atom force field[72] was used for the protein. The solvent was described by the TIP4P water model.[73] Electrostatic interactions were calculated explicitly at a distance <1 nm, and long-range electrostatic interactions were calculated at every step by particle-mesh Ewald summation.[74] Lennard–Jones interactions were calculated with a cutoff of 1 nm. All bonds were constrained by using the LINCS algorithm,[75] allowing for an integration time step of 2 fs. The simulation temperature was kept constant by weakly ($\tau=0.1$ ps) coupling the lipids, protein, and solvent (water+counter-ions) separately to a temperature bath of 300 K. The pressure was kept constant by weakly coupling the system to a pressure bath of 1 bar with semi-isotropic pressure coupling. Prior to simulations, 500 conjugate gradient energy-minimization steps were performed, followed by 2 ns of restrained MD in which the protein atoms were restrained with a force constant of 1000 kJ mol^{-1} nm^{-2} to their initial position. Ions, lipids, and solvent were allowed to move freely during this 2 ns equilibration phase. The system was then subjected to 20 ns of unrestrained MD, during which coordinates were saved every 10 ps for analysis. Parts of models 2 and 7 started to unfold after several ns; therefore, simulations were stopped after 10 ns. Simulations for models 1 and 6 were extended to 60 and 100 ns, respectively. pK_a values for all titratable amino acid side chains within the models were calculated using PROPKA.[76] Residues at the N- and C-termini were considered as uncharged, as neither lie at the actual termini of the complete channel.

Drug docking

FlexX v. 3.0.2[77] and the GOLD evaluation v. 4.0.1[78] with standard parameters were used to analyze hERG interactions with (+)-cisapride, (S)-terfenadine, and MK-499. Drug coordinates were obtained from the PubChem structure database.[79] The starting geometries of the drugs were optimized with the Hartree–Fock, 6-31G* basis set, as implemented in Gaussian 03.[54b] Structures prior to MD simulations and minimized averaged structures (final 500 ps) were used as starting coordinates for docking. The highest-scoring docking poses for each model were stored, and results of different hERG models were compared.

Acknowledgements

We thank Ulrike Gerischer for carefully reading the manuscript. This work was funded by TI Pharma project D2-101 (A.S. and

U.Z.) and EC project HEALTH-F4-2007-201924, EDICT Consortium (S.J.W.).

Keywords: docking · hERG · model validation · molecular dynamics · molecular modeling

[1] P. Ju, G. Pages, R. P. Riek, P. C. Chen, A. M. Torres, P. S. Bansal, S. Kuyucak, P. W. Kuchel, J. I. Vandenberg, *J. Biol. Chem.* **2009**, *284*, 1000–1008.
[2] M. C. Sanguinetti, C. Jiang, M. E. Curran, M. T. Keating, *Cell* **1995**, *81*, 299–307.
[3] M. C. Trudeau, J. W. Warmke, B. Ganetzky, G. A. Robertson, *Science* **1995**, *269*, 92–95.
[4] G. N. Tseng, *J. Mol. Cell. Cardiol.* **2001**, *33*, 835–849.
[5] J. I. Vandenberg, B. D. Walker, T. J. Campbell, *Trends Pharmacol. Sci.* **2001**, *22*, 240–246.
[6] A. Arcangeli, L. Bianchi, A. Becchetti, L. Faravelli, M. Colonnello, E. Mini, M. Olivotto, E. Wanke, *J. Physiol.* **1995**, *489*, 455–471.
[7] M. E. Curran, I. Splawski, K. W. Timothy, G. M. Vincent, E. D. Green, M. T. Keating, *Cell* **1995**, *80*, 795–803.
[8] R. Pearlstein, R. Vaz, D. Rampe, *J. Med. Chem.* **2003**, *46*, 2017–2022.
[9] C. Jamieson, E. M. Moir, Z. Rankovic, G. Wishart, *J. Med. Chem.* **2006**, *49*, 5029–5046.
[10] J. S. Mitcheson, J. Chen, M. Lin, C. Culberson, M. C. Sanguinetti, *Proc. Natl. Acad. Sci. USA* **2000**, *97*, 12329–12333.
[11] R. Farid, T. Day, R. A. Friesner, R. A. Pearlstein, *Bioorg. Med. Chem.* **2006**, *14*, 3160–3173.
[12] M. Masetti, A. Cavalli, M. Recanatini, *J. Comput. Chem.* **2008**, *29*, 795–808.
[13] P. J. Stansfeld, P. Gedeck, M. Gosling, B. Cox, J. S. Mitcheson, M. J. Sutcliffe, *Proteins* **2007**, *68*, 568–580.
[14] G. N. Tseng, K. D. Sonawane, Y. V. Korolkova, M. Zhang, J. Liu, E. V. Grishin, H. R. Guy, *Biophys. J.* **2007**, *92*, 3524–3540.
[15] R. Rajamani, B. A. Tounge, J. Li, C. H. Reynolds, *Bioorg. Med. Chem. Lett.* **2005**, *15*, 1737–1741.
[16] F. Österberg, J. Åqvist, *FEBS Lett.* **2005**, *579*, 2939–2944.
[17] R. A. Pearlstein, R. J. Vaz, J. Kang, X. L. Chen, M. Preobrazhenskaya, A. E. Shchekotikhin, A. M. Korolev, L. N. Lysenkova, O. V. Miroshnikova, J. Hendrix, D. Rampe, *Bioorg. Med. Chem. Lett.* **2003**, *13*, 1829–1835.
[18] C. Chothia, A. M. Lesk, *EMBO J.* **1986**, *5*, 823–826.
[19] C. Sander, R. Schneider, *Proteins* **1991**, *9*, 56–68.
[20] D. A. Doyle, J. M. Cabral, R. A. Pfuetzner, A. Kuo, J. M. Gulbis, S. L. Cohen, B. T. Chait, R. MacKinnon, *Science* **1998**, *280*, 69–77.
[21] Y. Jiang, A. Lee, J. Chen, M. Cadene, B. T. Chait, R. MacKinnon, *Nature* **2002**, *417*, 515–522.
[22] Y. Jiang, A. Lee, J. Chen, V. Ruta, M. Cadene, B. T. Chait, R. MacKinnon, *Nature* **2003**, *423*, 33–41.
[23] S. B. Long, E. B. Campbell, R. MacKinnon, *Science* **2005**, *309*, 897–903.
[24] S. B. Long, X. Tao, E. B. Campbell, R. MacKinnon, *Nature* **2007**, *450*, 376–382.
[25] K. Kamiya, R. Niwa, M. Morishima, H. Honjo, M. C. Sanguinetti, *J. Pharmacol. Sci.* **2008**, *108*, 301–307.
[26] G. M. Clayton, S. Altieri, L. Heginbotham, V. M. Unger, J. H. Morais-Cabral, *Proc. Natl. Acad. Sci. USA* **2008**, *105*, 1511–1515.
[27] H. P. Larsson, F. Elinder, *Neuron* **2000**, *27*, 573–583.
[28] J. Liu, M. Zhang, M. Jiang, G. N. Tseng, *J. Gen. Physiol.* **2002**, *120*, 723–737.
[29] K. Hayashi, M. Shimizu, H. Ino, M. Yamaguchi, H. Mabuchi, N. Hoshi, H. Higashida, *Cardiovasc. Res.* **2002**, *54*, 67–76.
[30] M. Fasnacht, J. Zhu, B. Honig, *Protein Sci.* **2007**, *16*, 1557–1568.
[31] J. Chen, G. Seebohm, M. C. Sanguinetti, *Proc. Natl. Acad. Sci. USA* **2002**, *99*, 12461–12466.
[32] G. Klement, J. Nilsson, P. Arnhem, F. Elinder, *Biophys. J.* **2008**, *94*, 3014–3022.
[33] Y. N. Imai, S. Ryu, S. Oiki, *J. Med. Chem.* **2009**, *52*, 1630–1638.
[34] J. Karczewski, J. Wang, S. A. Kane, L. Kiss, K. S. Koblan, J. C. Culberson, R. H. Spencer, *Biochem. Pharmacol.* **2009**, *77*, 1602–1611.
[35] Y. Shafrir, S. R. Durell, H. R. Guy, *Biophys. J.* **2008**, *95*, 3650–3662.
[36] T. Kuner, P. H. Seeburg, H. R. Guy, *Trends Neurosci.* **2003**, *26*, 27–32.
[37] S. K. Burley, G. A. Petsko, *Science* **1985**, *229*, 23–28.
[38] S. Persson, J. A. Killian, G. Lindblom, *Biophys. J.* **1998**, *75*, 1365–1371.
[39] I. T. Arkin, A. T. Brunger, *Biochim. Biophys. Acta* **1998**, *1429*, 113–128.
[40] C. Domene, S. Vemparala, M. L. Klein, C. Vénien-Bryan, D. A. Doyle, *Biophys. J.* **2006**, *90*, L01–L03.
[41] P. Braun, G. von Heijne, *Biochemistry* **1999**, *38*, 9778–9782.
[42] Y. Liu, D. M. Engelman, M. Gerstein, *Genome Biol.* **2002**, *3*, research0054.1–0054.12.
[43] S. C. Li, C. M. Deber, *Nat. Struct. Biol.* **1994**, *1*, 368–373.
[44] R. J. Law, C. Capener, M. Baaden, P. J. Bond, J. Campbell, G. Patargias, Y. Arinaminpathy, M. S. Sansom, *J. Mol. Graphics Modell.* **2005**, *24*, 157–165.
[45] L. R. Forrest, M. S. Sansom, *Curr. Opin. Struct. Biol.* **2000**, *10*, 174–181.
[46] A. Ivetac, M. S. Sansom, *Eur. Biophys. J.* **2008**, *37*, 403–409.
[47] J. F. Taly, A. Marin, J. F. Gibrat, *BMC Bioinf.* **2008**, *9*, 6.
[48] A. Stary, Y. Shafrir, S. Hering, P. Wolschann, H. R. Guy, *Channels* **2008**, *2*, 210–215.
[49] J. P. Lees-Miller, J. O. Subbotina, J. Guo, V. Yarov-Yarovoy, S. Y. Noskov, H. J. Duff, *Biophys. J.* **2009**, *96*, 3600–3610.
[50] T. Myokai, S. Ryu, H. Shimizu, S. Oiki, *Mol. Pharmacol.* **2008**, *73*, 1643–1651.
[51] D. Fernandez, A. Ghanta, G. W. Kauffman, M. C. Sanguinetti, *J. Biol. Chem.* **2004**, *279*, 10120–10127.
[52] E. Ficker, W. Jarolimek, J. Kiehn, A. Baumann, A. M. Brown, *Circ. Res.* **1998**, *82*, 368–395.
[53] O. Casis, S. P. Olesen, M. C. Sanguinetti, *Mol. Pharmacol.* **2006**, *69*, 658–665.
[54] a) M. Perry, F. B. Sachse, M. C. Sanguinetti, *Proc. Natl. Acad. Sci. USA* **2007**, *104*, 13827–13832; b) M. J. Frisch, Gaussian 03 (Revision C.02), Gaussian Inc., Wallingford, CT (USA), **2004**.
[55] M. A. Marti-Renom, A. Stuart, A. Fiser, R. Sánchez, F. Melo, A. Sali, *Annu. Rev. Biophys. Biomol. Struct.* **2000**, *29*, 291–325.
[56] S. Y. Lee, A. Lee, J. Chen, R. MacKinnon, *Proc. Natl. Acad. Sci. USA* **2005**, *102*, 15441–15446.
[57] J. U. Bowie, R. Lüthy, D. Eisenberg, *Science* **1991**, *253*, 164–170.
[58] R. A. Laskowski, M. W. MacArthur, D. S. Moss, J. M. Thornton, *J. Appl. Crystallogr.* **1993**, *26*, 283–291.
[59] R. W. Hooft, G. Vriend, C. Sander, E. E. Abola, *Nature* **1996**, *381*, 272–272.
[60] M. Wiederstein, M. J. Sippl, *Nucleic Acids Res.* **2007**, *35*, W407–W410.
[61] P. Larsson, B. Wallner, E. Lindahl, A. Elofsson, *Protein Sci.* **2008**, *17*, 990–1002.
[62] H. Li, Y. Zhou, *J. Bioinf. Comput. Biol.* **2005**, *3*, 1151–1170.
[63] L. J. McGuffin, *Bioinformatics* **2008**, *24*, 586–587.
[64] C. S. Pettitt, L. J. McGuffin, D. T. Jones, *Bioinformatics* **2005**, *21*, 3509–3515.
[65] L. J. McGuffin, *BMC Bioinf.* **2007**, *8*, 345.
[66] W. Kabsch, C. Sander, *Biopolymers* **1983**, *22*, 2577–2637.
[67] D. T. Jones, *J. Mol. Biol.* **1999**, *292*, 195–202.
[68] E. Lindahl, B. Hess, D. van der Spoel, *J. Mol. Model.* **2001**, *7*, 306–317.
[69] J. D. Faraldo-Gómez, G. R. Smith, M. S. Sansom, *Eur. Biophys. J.* **2002**, *31*, 217–227.
[70] J. Åqvist, V. Luzhkov, *Nature* **2000**, *404*, 881–884.
[71] O. Berger, O. Edholm, F. Jähnig, *Biophys. J.* **1997**, *72*, 2002–2013.
[72] W. L. Jorgensen, D. S. Maxwell, J. Tirado-Rives, *J. Am. Chem. Soc.* **1996**, *118*, 11225–11236.
[73] W. L. Jorgensen, J. Chandrasekhar, J. D. Madura, R. W. Impey, M. L. Klein, *J. Chem. Phys.* **1983**, *79*, 926–935.
[74] T. Darden, D. York, L. Pedersen, *J. Chem. Phys.* **1993**, *98*, 10089–10092.
[75] B. Hess, H. Bekker, H. J. C. Berendsen, J. G. E. M. Fraaije, *J. Comput. Chem.* **1997**, *18*, 1463–1472.
[76] H. Li, A. D. Robertson, J. H. Jensen, *Proteins* **2005**, *61*, 704–721.
[77] M. Rarey, FlexX 3.0.2, BioSolveIT GmbH, St. Augustin (Germany) **2008**.
[78] G. Jones P. Willett, R. C. Glen, *J. Mol. Biol.* **1995**, *245*, 43–53.
[79] PubChem: http://pubchem.ncbi.nlm.nih.gov (accessed January 7, 2010).

Received: November 10, 2009
Revised: December 19, 2009
Published online on January 26, 2010

5.3 Paper #3

Different Inward and Outward Conduction Mechanisms in Na$_V$Ms Suggested by Molecular Dynamics Simulations

Song Ke, E. N. Timin, Anna Stary-Weinzinger*

Department of Pharmacology and Toxicology, University of Vienna, Vienna, Austria

Abstract

Rapid and selective ion transport is essential for the generation and regulation of electrical signaling pathways in living organisms. Here, we use molecular dynamics (MD) simulations with an applied membrane potential to investigate the ion flux of bacterial sodium channel Na$_V$Ms. 5.9 μs simulations with 500 mM NaCl suggest different mechanisms for inward and outward flux. The predicted inward conductance rate of ~27±3 pS, agrees with experiment. The estimated outward conductance rate is 15±3 pS, which is considerably lower. Comparing inward and outward flux, the mean ion dwell time in the selectivity filter (SF) is prolonged from 13.5±0.6 ns to 20.1±1.1 ns. Analysis of the Na$^+$ distribution revealed distinct patterns for influx and efflux events. In 32.0±5.9% of the simulation time, the E53 side chains adopted a flipped conformation during outward conduction, whereas this conformational change was rarely observed (2.7±0.5%) during influx. Further, simulations with dihedral restraints revealed that influx is less affected by the E53 conformational flexibility. In contrast, during outward conduction, our simulations indicate that the flipped E53 conformation provides direct coordination for Na$^+$. The free energy profile (potential of mean force calculations) indicates that this conformational change lowers the putative barriers between sites S$_{CEN}$ and S$_{HFS}$ during outward conduction. We hypothesize that during an action potential, the increased Na$^+$ outward transition propensities at depolarizing potentials might increase the probability of E53 conformational changes in the SF. Subsequently, this might be a first step towards initiating slow inactivation.

Citation: Ke S, Timin EN, Stary-Weinzinger A (2014) Different Inward and Outward Conduction Mechanisms in Na$_V$Ms Suggested by Molecular Dynamics Simulations. PLoS Comput Biol 10(7): e1003746. doi:10.1371/journal.pcbi.1003746

Editor: Emad Tajkhorshid, University of Illinois, United States of America

Received December 17, 2013; Accepted June 13, 2014; Published July 31, 2014

Copyright: © 2014 Ke et al. This is an open-access article distributed under the terms of the Creative Commons Attribution License, which permits unrestricted use, distribution, and reproduction in any medium, provided the original author and source are credited.

Funding: This work was supported by The Austrian Science Fund (FWF) Project W1232. The funders had no role in study design, data collection and analysis, decision to publish, or preparation of the manuscript.

Competing Interests: The authors have declared that no competing interests exist.

* Email: anna.stary@univie.ac.at

Introduction

Na$^+$ flux through voltage gated sodium channels (Na$_V$) is crucial for initiating action potentials in the membranes of electrically excitable cells. They mediate a variety of biological functions such as muscle contraction, propagation of nerve impulses, release of hormones and many more [1]. As a consequence, mutations in Na$_V$ channels lead to a variety of channelopathies, such as congenital epilepsy, cardiac arrhythmias or chronic pain [2,3].

Recently, homotetrameric crystal structures of several bacterial Na$_V$ channels were successfully resolved [4–10], providing a tremendous opportunity to investigate the structure and function of these channels on the atomistic level. They are composed of four membrane spanning subunits and contain six transmembrane (TM) helices per subunit. Helices S1 to S4 form the voltage sensing module. Helices S5, P1 segments, the selectivity filter (SF) region, P2 segments and S6 helices, lining the inner pore cavity, form the pore module. The SF of most bacterial channels contains the amino acid sequence TLESW. The four glutamic acid side chains [11] form a high field strength binding site (HFS) [12] which is essential for ion selectivity. In eukaryotic sodium channels, this site consists of the amino acids motif DEKA.

The molecular mechanisms underlying ion conduction and selectivity in Na$_V$ are beginning to emerge. Computational methods, particularly molecular dynamics (MD) simulations are extensively adopted to address these questions [13–23].

As reviewed recently [24], sodium ions were illustrated to spontaneously traverse the SF into the cavity with energy barriers between ~2–5 kcal/mol. Compared to potassium coordination in K$_V$ channels, sodium ions partially or fully preserve their first hydration shells [13,15–17,20,22]. A loosely coupled knock-on mechanism with an average ion occupancy around two in the SF was predicted during ion conduction [14,15,21,22]. The incoming ion repulses the present ion out of the SF. The wide radius (≥9 Å) of the SF enables double occupancy of ions at the same level [20,21]. Further, Na$^+$ vs. K$^+$ [15,16,20,20–22] and Na$^+$ vs. Ca^{2+} [17,20] discrimination studies were carried out. These studies revealed higher energy barriers in the SF for K$^+$ and Ca^{2+} compared to Na$^+$. Subsequently, non-equilibrium simulations were performed to investigate conduction under applied membrane potentials and to study kinetics [21,22]. The estimated inward conductance rate successfully reproduced electrophysiology data [22].

A recent study by Chakrabarti et al. [23], suggested that conformational changes at the EEEE motif (corresponding to E177 in Na$_V$Ab) might play an important role in ion conduction. However, this observation was not reported in other simulations, except for simulations using Ca^{2+} as a charge carrier [17]. In K$^+$ channels, subtle structural changes in the SF, involving rotations around a highly conserved glycine residue result in different non-conductive conformations [25,26]. This regulation of ion flow by conformational changes of the selectivity filter is termed C-type inactivation.

Author Summary

Voltage gated sodium channels are essential components of living cell membranes. They regulate the cell potential by facilitating permeation of ions across the membrane. In the past decades, studies revealed that the bacterial selectivity filter (SF) exhibits a constricted architecture lined with electronegative carboxyl oxygens of four glutamic acid side chains (EEEE motif), which repulse anions but attract Na^+ ions. Crystal structures enable the investigation of structural dynamics with computational methods. Ion selectivity and conduction mechanisms between Na^+, K^+ and Ca^{2+} are progressively elucidated by molecular dynamics simulations and free energy calculations. The structural dynamics of the protein, especially the flexibility of SF and its fundamental role in kinetics underpinning ion selectivity, conduction and channel gating are less well understood. To shed light on this question, we use computational simulations to simulate ion conduction with membrane potentials. Our results suggest different dynamical behaviors of the EEEE locus and distinct ion distribution patterns in the SF with respect to permeating directionalities. These findings indicate a novel mechanism in differentiating reciprocal transitions of ion flow, preventing large sodium efflux during action potential initiation and may further suggest that increased flipping propensities at depolarizing potentials, might initially trigger channel slow inactivation.

It is not clear to which extend structural changes at the EEEE locus in Na_v channels are crucial for conductance and inactivation in Na_v channels.

To investigate these issues, we conducted MD simulations using the open conformation of the bacterial sodium channel homologue Na_vMs (Magnetococcus sp. (strain MC-1)) [7] pore domain focusing on the structural changes of the SF during inward and outward conduction.

Results

Single channel conductance

The four-fold symmetrical structure of Na_vMs (pdb identifier: 4F4L) was generated using chain A (splayed outward by $25°$ rotation about its Ψ-bond at position T84), which creates an open pore with a diameter of ~ 14 Å [7]. As described by Ulmschneider et al. [22], a harmonic restraint was exerted on the S5 and S6 TM helices to keep the gate in the open conformation throughout simulations. This structure was then embedded into a POPC lipid patch and duplicated in the Z direction (pore axis). A constant charge imbalance of four elementary charges (4 e) across each lipid bilayer between the central electrolyte bath and the two outer ones was maintained during simulation (supplementary movie S1) [27].

Four times 500 ns double-patch MD simulations with 500 mM NaCl were performed, with the first 100 ns treated as equilibration. Figure 1 shows the cumulative ion conducting events from MD simulations with depolarized and hyperpolarized membrane potentials of ΔV: 565 ± 126 mV (Figure 2A). The estimated sodium current in the inward direction is $\gamma = 27\pm3$ pS. This value agrees with previously observed single channel conductance measurements ($\gamma\sim33$ pS) [22]. Our double bilayer simulations enabled us to further estimate outward conduction, which amounts to $\gamma = 15\pm3$ pS. Interestingly, this process is distinguishably slower than inward ion flux (P<0.01, N = 4).

Ion distribution patterns and kinetics

To explore the underlying differences between inward and outward ion permeation, we plotted the ion probability density map across the pore region from all four simulations. Several favorable ion-interacting sites (S_{EX}, S_{HFS}, S_{CEN} and S_{IN}) from periplasm to cytoplasm were assigned as proposed previously by Payandeh et al. [4]. The ion-interacting sites across the SF were determined by measuring the axial distance (Z axis) along certain atoms from −5.00 to 10.25 Å as shown in Figure 3. Side chain oxygens of S54 from all four chains were taken as the origin (Z = 0.0 Å) of the SF. Additionally, in two previous studies, a site with an energy barrier ($\sim$2 kcal/mol) distinguishing between site S_{HFS} and S_{CEN} was identified [17,22]. In this study, we refer to this site as "S_{BAR}", indicating this barrier (2.75≤Z<4.75 Å).

During influx (Figure 4A and B), short-lived Na^+ binding at site S_{EX} was observed (2.6±0.5 ns) in an asymmetrical manner. S_{HFS} is the dominant site with the highest ion density. At this site, ions tended to be directly coordinated with side chain oxygens of E53 and S54 in an off-axis manner. Additionally, a less densely populated configuration was observed in the center of this site consistent with previous studies [21,24]. Moving inward from site S_{HFS}, ions further translocated transiently via site S_{BAR} (1.5±0.3 ns) to site S_{CEN} (3.4±0.3 ns). Subsequently, ions reciprocally traversed between sites S_{CEN} and S_{IN}. These results are in good agreement with previous simulation studies [14,15,17,21,22].

Our simulations revealed a distinct ion distribution pattern for efflux compared to influx as shown in Figure 4C and D. After entering into site S_{IN} from the cytosol, ions mainly populated sites S_{CEN} and S_{BAR} with extended dwell times compared to inward conduction, (S_{CEN}: 11.9±1.1 ns vs. 3.4±0.3 ns; S_{BAR}: 8.2±0.9 ns vs. 1.5±0.3 ns) suggesting a putative barrier for efflux between sites S_{BAR} and S_{HFS}. Additionally, during efflux, Na^+ ions tended to traverse in an on-axis manner through the filter.

Structural dynamics of glutamic acid

Conformational isomerization of the E53 side chains has been reported previously [17,23]. Generally, the glutamic acid side chain might adopt two main conformations (Figure 5A and B): inward-facing (χ_2 angle $\sim60°$, flipped) and outward-facing (χ_2 angle $\sim290°$, non-flipped). In our simulations, during influx, flipping events were observed only in 2.7±0.5% of the simulation time, thus the E53 side chain mainly adopted a non-flipped conformation. In contrast, during efflux 32.0±5.9% flipping events were observed (P value = 0.015, see Figure 5C). A more detailed investigation of this flipping events revealed that 80% of these changes occurred in only one of the four glutamic acid side chains ("one-flip") (Figure 5D).

To investigate the influence of the presence of Na^+ ions on E53 side chain dynamics, three repeated simulations without ions in the SF ("no salt") were performed. Irrespective of the directionality of the applied potentials, the flip probability is less than 0.6% in all simulations (Figure 5C). This indicates, that a depolarizing potential per se does not significantly influence the number of E53 flipping events. This suggests that the combination of local positive charge carried by outward Na^+ flux in the SF especially at sites S_{HFS} and S_{BAR} and the outward attracting membrane potential might collectively induce the rotation of the χ_2 angle from $\sim290°$ to $\sim60°$.

Ionic binding modes and conduction mechanism during inward conduction

A detailed investigation of the ionic binding modes and their relations to free energy profiles enabled us to describe mechanisms

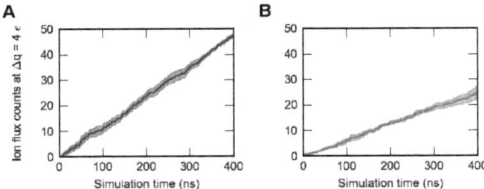

Figure 1. Ion flux in double bilayer simulations without dihedral restraints on E53 (n = 4). A) Average ion flux count of inward conduction through the SF over time (color: blue). B) Average ion flux count of outward conduction through the SF over time (color: red). Error estimations shown in the figure are S.E.M.
doi:10.1371/journal.pcbi.1003746.g001

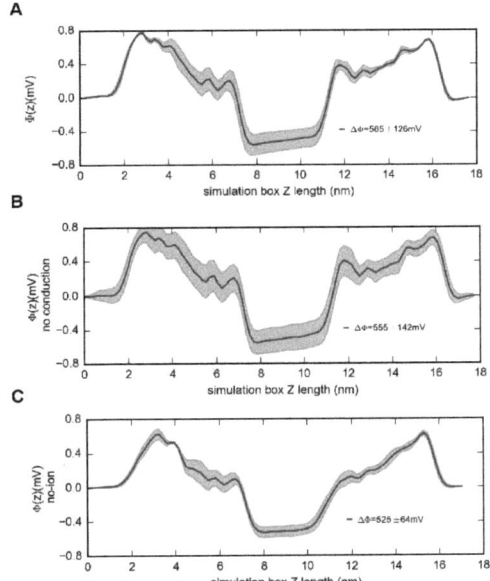

Figure 2. Transmembrane voltages comparison. A) Electrostatic potential across the simulation box calculated from the simulations without dihedral restraints on E53. B) Electrostatic potential across the simulation box calculated from the snapshots (t>5 ns) extracted from the simulation in Figure 2A without ion conduction events in neither inward nor outward directions. C) Electrostatic potential across the simulation box calculated from the "no salt" simulations. (Error estimations shown in the figure are S.D.).
doi:10.1371/journal.pcbi.1003746.g002

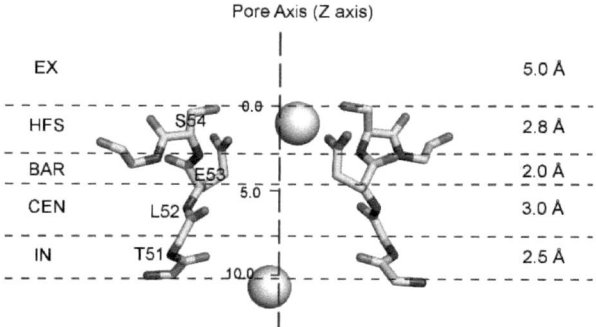

Figure 3. Annotations of the SF ion interacting site. SF backbone atoms are shown as blue sticks, E53 and S54 side chain atoms are also shown to determine the site S_{HFS} (only two opposite subunits are shown for clarity). In addition, sodium ions are depicted with yellow spheres. Sites S_{EX} ($-5.00 \leq Z < 0.00$ Å), S_{HFS} ($0.00 \leq Z < 2.75$ Å), S_{BAR} ($2.75 \leq Z < 4.75$ Å), site S_{CEN} ($4.75 \leq Z < 7.75$ Å) and site S_{IN} ($7.75 \leq Z < 10.25$ Å) and the length of each site were annotated at the lateral sides of the figure.
doi:10.1371/journal.pcbi.1003746.g003

regarding different conducting directionalities (Figure 6 and 7). During inward conduction, the largest barrier in the SF occurs between sites S_{HFS} and S_{BAR} which amounts to 2.1 kcal/mol (Figure 6B). At site S_{HFS}, the probe ions (yellow) mainly distributed in an off-axis manner, the first coupling Na$^+$ ions (blue) may occupy site $S_{CEN (IN)}$ and there existed a second binding site for coupling ions at site S_{EX} (Figure 6A, II). Subsequently, the probe ions distributed in the middle of channel axis when traversing the short lived site S_{BAR}, with the other two coupling ions populating sites S_{EX} and $S_{IN (CAV)}$ respectively (Figure 6A, III). The probe ions then occupied site S_{CEN} in both on-axis and off-axis manners, other coupling ions in the SF were distributed mainly at sites S_{IN} and S_{EX}. Only a few coupled ions occupied sites S_{HFS} and S_{BAR} (Figure 6A, IV).

In these ionic binding modes, under hyperpolarized potential, the coupling ions in the SF generally demonstrated a loosely coupled knock-on mechanism with only a few of them present in the adjacent binding sites to the probe ions (Figure 6A, II–IV). This is in agreement with a study by [21,22], where it was shown that during inward conduction the ions displayed a combination of mono-ionic and multi-ionic mechanism with an overall occupancy of 1.8 ions in the pore region.

The flipping probability analysis indicates that the conformational changes of the E53 side chains play a minor role for ion inward conduction as shown in Figure 6A, II'–IV'.

Ionic binding modes and conduction mechanism during outward conduction

Compared to inward permeation, the maximum energy barrier during outward conduction amounted to 2.3 kcal/mol between sites S_{BAR} and S_{HFS}. It is interesting that the free energy difference between sites S_{CEN} and S_{HFS} is 2.2 kcal/mol, which is close to the largest energy barrier (Figure 7B). In addition, the ionic binding modes demonstrate a distinct conduction mechanism compared to Na$^+$ influx. Traversing outward from the cavity, ions at site S_{IN} were tightly coupled with ions at site S_{BAR} (Figure 7A, III and V) corresponding to two energy wells in Figure 7B, III and V). When probe ions located at site S_{CEN}, the coupling ions distributed in the upper part of the cavity (Figure 7A, IV) which corresponds to the energy well at site S_{CEN} (Figure 7B, IV). Generally, the translocation of probe ions from the cytoplasm to site S_{BAR} is readily stepwise by a tight knock-off mechanism without significant energy barriers. At all three energy wells (Figure 7B III–V) the E53 side chains maintained non-flipped conformations.

When probe ions faced the energy barrier at site S_{BAR}, a delicate tightly-coupled "knock-off" conducting mechanism occurred. Initially, E53 started to flip and one of the carboxyl oxygens started to coordinate the probe ions (Figure 7A, III' and S5, B). Compared to ions located in the close energy wells (Figure 7A, III), the probe ions were meanwhile expulsed by the outward movements of approaching coupling ions at site S_{IN} (Figure 7A, III'). If this knock-off mechanism was successful, the probe ions would then migrate to site S_{HFS}, as a result, the coupling ions would move outward to site S_{CEN} simultaneous (Figure 7A, II' and IV'). At this time, two carboxyl oxygens of the flipped E53 side chain tended to coordinate with the probe ions and coupling ions respectively (Figure 7A, II' and IV' and S5, B). Afterwards, ions left site S_{HFS} promptly (t = 2.6 ns) into the periplasm via site S_{EX}.

If the attempt to overcome the barrier failed, the aforementioned mechanism was easily reversed, the probe ions and coupling ions occupied the two stable energy wells at sites S_{BAR} with the coupling ions at site S_{IN} (Figure 7A, III and V) and site S_{CEN} with the coupling ions in the cavity (Figure 7A, IV) again. That is the reason why ions stayed longer in sites S_{BAR} and S_{CEN}.

One the one hand, larger Pi values (flip inducing probability of number of probe ions, see methods for details) values of sites S_{BAR}, S_{CEN} and S_{HFS} indicated the flipped conformations of E53 were crucial (Pi>90%) in overcoming the dual energy barriers between S_{CEN}, S_{BAR} and S_{HFS}. On the other hand, smaller Pt values

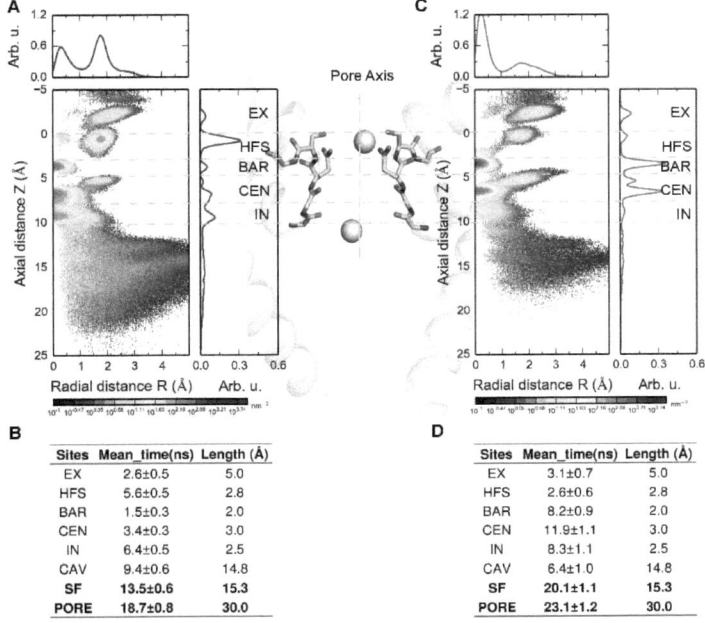

Figure 4. Ion distribution probability density map. A) Ion distribution of inward simulations labeled with respective interacting sites [4], lower left: 2-D (axial and radial distribution) density; lower right: 1-D axial distribution along the pore axis; upper left: 1-D radial distribution from the center of the pore axis. B) Inward dwell time in respective interacting sites (158 ion conduction events from four simulations were taken for analysis). C) Ion distribution of outward simulations labeled with respective interacting sites [4], lower left: 2-D (axial and radial distribution) density; lower right: 1-D axial distribution along the pore axis; upper left: 1-D radial distribution from the center of pore axis. D) Outward dwell time in respective interacting sites (79 ion conduction events from four simulations were taken for analysis).
doi:10.1371/journal.pcbi.1003746.g004

(flipping time probability for all probe ions, see methods for details) values indicated that the flipping events were easily reversible. Because of these flipping events, the major ion distribution for outward conduction is limited to the center of the channel axis during translocation within the SF.

E53 conformation determines efflux rate

To further explore the correlation between flux directionality and E53 conformation, we performed two sets of inward and outward conduction simulations (four times 300 ns) with dihedral restraints to maintain "non-flip" and "one-flip" configurations during sampling. The influx rate was independent of the E53 conformations as shown in Figure 8A. This observation disagrees with recent data from Chakrabarti et al. [23] on the Na$_V$Ab channel. The outward conduction with "one-flip" simulation displayed an increased efflux rate compared to simulations without dihedral restraints on E53, where the flipping events would be reversible when conducting ions (Figures S1, S2, S3, S4). Interestingly, if E53 was restrained to a "non-flip" configuration, sodium ions translocation slowed down (Figure 8B). These results suggest a clear influence of filter dynamics on the efflux rate.

Comparison of the free energy profiles from outward simulations of these three types of configurations revealed that the largest energy barrier of the "non-flip" simulations is increased from 2.3 kcal/mol (non-restraint) to 3.4 kcal/mol (Figure 9A and C). The lowest energy well at site S_{BAR} was also replaced by site S_{CEN}. The energy profile of "one-flip" simulations indicated that the energy barrier between sites S_{CEN} and S_{HFS} was diminished,

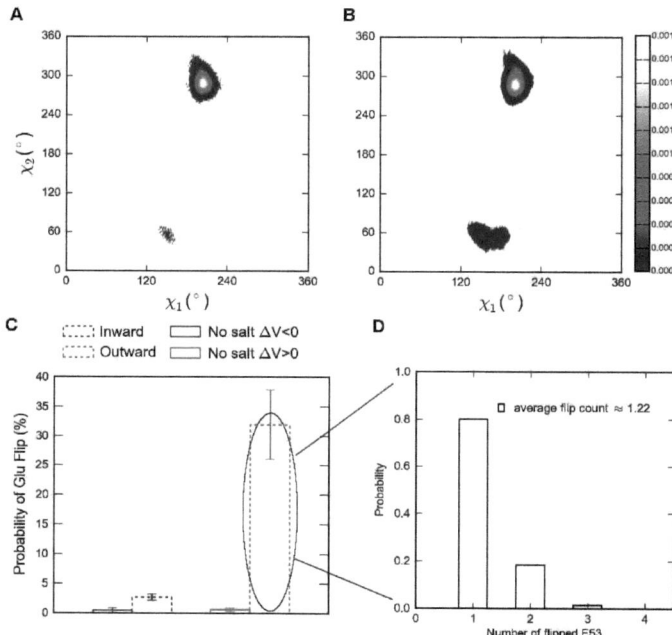

Figure 5. E53 conformational changes. A) χ_1-χ_2 angle distribution of the inward conductance simulations. B) χ_1-χ_2 angle distribution of the outward conductance simulations. C) The probability of flipping events of E53 side chains, the inward conduction is depicted as blue dotted histogram. The outward conduction is depicted as red dotted histogram (a flip is defined when at least one out of four subunits was shifted to flipping state in an analyzing window over time from four simulations); Control simulations with ions not in the SF ("no salt") at hyperpolarized and depolarized membrane potentials are depicted as blue and red solid histograms, respectively. Error estimations shown in the figure are S.E.M. D) The distribution of E53 flipping numbers from all flipping events during outward conduction.
doi:10.1371/journal.pcbi.1003746.g005

although the original energy barrier increased slightly by 0.3 kcal/mol (Figure 9A and B). Therefore, the outward conduction would be more straightforward without reversible backward translocation compared to the simulations without the dihedral restraints from a kinetic point of view.

Discussion

The mechanism of ion conduction and selectivity of bacterial voltage gated sodium channels is gradually emerging. The inverted tepee shape architecture of the SF lined with TLESW sequence enables sodium influx at the diffusion rate. The glutamic acid side chains are responsible for recruiting Na$^+$ ions from the outer vestibule. Ions will then translocate via sites S_{HFS}, S_{BAR}, S_{CEN} and S_{IN} spontaneously to complete a conduction event [21–23]. Details of conduction are only partially understood. Large conformational changes of the glutamic acid side chains were described recently [23]. However, its role for conduction is still under discussion [24].

To gain further insights into these questions, we performed MD simulations to compare the different binding patterns and characterize the structural dynamics of glutamic acids during ion permeation. Double bilayer simulations with the open Na$_V$Ms structure enabled us to investigate influx and efflux separately. The calculated inward conductance rate is in good agreement with a previously reported experiment and computational data [22]. The

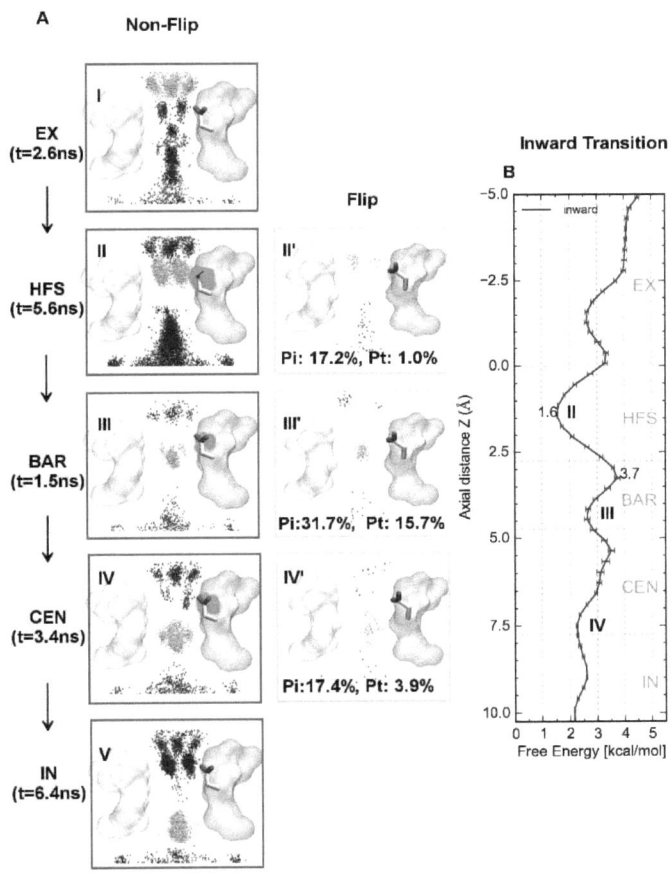

Figure 6. Inward ionic binding modes and PMF. A) Overlay of the conformational space of the probe ions (yellow), coupling sodium ions (blue) and E53 carboxyl oxygen distributions (transparent spheres with different intensity) for different ionic binding modes at different interaction sites. A snapshot of a typical E53 sidechain conformation is shown as stick representation (red, carboxyl oxygens; yellow, sidechain carbons). I–V are non-flip binding modes and II'–IV' are flipping binding modes for sites S_{HFS}, S_{BAR} and S_{CEN}. Arrows indicate the direction of ion conduction; B) 1-D PMF of inward conduction in SF region, the largest free energy barrier is labeled by a terminal peak and well, corresponding positions of typical binding modes on the energy profile are labeled. Error estimations shown in the figure are S.E.M.
doi:10.1371/journal.pcbi.1003746.g006

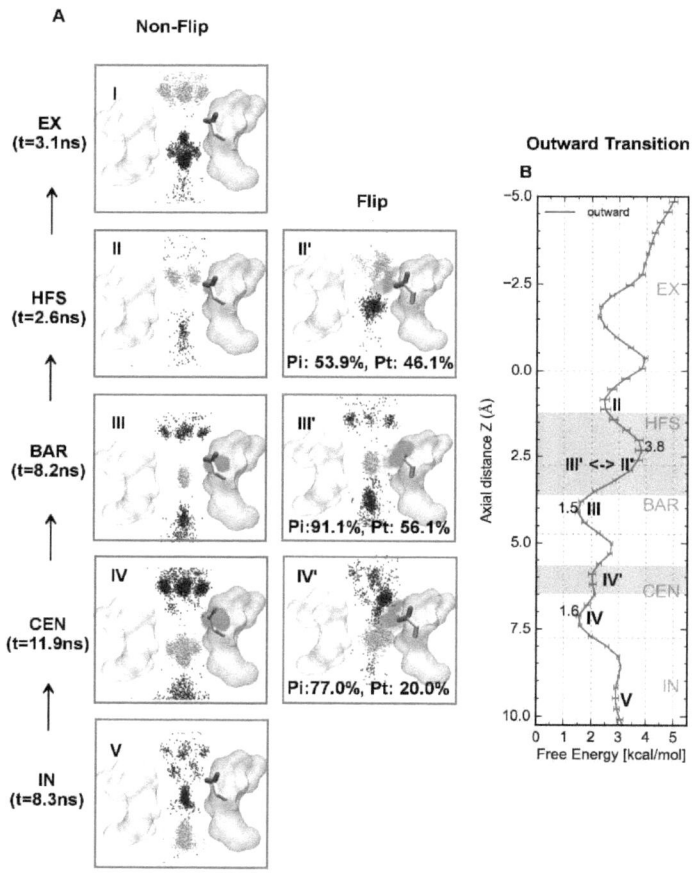

Figure 7. Outward ionic binding modes and PMF. A) Overlay of the conformational space of the probe ions (yellow), coupling sodium ions (blue) and E53 carboxyl oxygen distributions (transparent spheres with different intensity) for different ionic binding modes at different interaction sites. A snapshot of a typical E53 sidechain conformation is shown as stick representation (red, carboxyl oxygens; yellow, sidechain carbons). I–V are non-flip binding modes and II'–IV' are flipping binding modes for sites S_{HFS}, S_{BAR} and S_{CEN}. Arrows indicate the direction of ion conduction; B) 1-D PMF of outward conduction in SF region, the largest free energy barriers are labeled by terminal peaks and wells, corresponding positions of typical binding modes on the energy profile are labeled. Error estimations shown in the figure are S.E.M.
doi:10.1371/journal.pcbi.1003746.g007

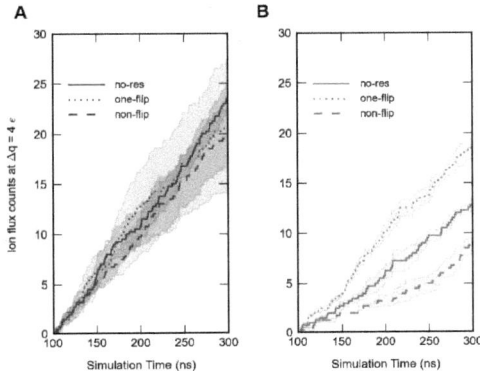

Figure 8. Influences of different flipping states at $\Delta q = 4e$ (n = 4). A) Average ion flux counts through the SF over time of inward conduction (color; blue). The simulations without restraints of E53 are depicted as solid line, the ones with the "one-flip" restraints are shown as dotted line and the ones with the "non-flip" restraints are shown as dashed line. B) Average ion flux count through the SF over time of outward conduction (color; red). Error estimations shown in the figure are S.E.M.
doi:10.1371/journal.pcbi.1003746.g008

estimated outward conductance rate obtained from MD simulations is predicted to be markedly lower than inward permeation (15 ± 3 pS vs 27 ± 3 pS, Figure 1). Ion translocation between sites S_{BAR} and S_{HFS} is substantially prolonged (8.2 ± 0.9 ns vs 1.5 ± 0.3 ns, Figure 4) during Na^+ efflux. From the energetic point of view, this would imply a potential barrier. This agrees with previous two-ion free energy calculation studies, revealing a higher energy barrier in this region for outward current compared to inward conduction (ΔG: 4.6 kcal/mol vs 0.4 kcal/mol, Stock et al. [21]; ΔG: 3.5±0.5 kcal/mol and 2.4±0.3 kcal/mol, Furini and Domene [15]). In our studies, this barrier is also higher for outward conduction (ΔG: 2.3 kcal/mol vs. 2.1 kcal/mol).

In agreement with previous studies [21,22] during inward conduction, our simulations revealed that ion translocations in the SF generally involve a loosely coupled knock on mechanism with an average ion occupancy of 1.8 (Figure 6).

A possible outward conduction mechanism was described by Stock et al., [21] using a "fully activated-open" Na_VAb channel structure [28]. They have found a third ion denoted k, directly coupling with the probe ions triggering outward conduction by a "nudging" collision effect. Similar results were obtained in our study, which shows that the coupling ions directly couple with the probe ions by a tight "knock-off" mechanism. Moreover, our simulations further elucidated that this "knock-off" mechanism is highly dependent on the conformational isomerization of the glutamic acid side chains in the SF. In other words, to overcome the energy barriers of outward conduction, at least one of the glutamic acid side chains has to be flipped to an inward facing conformation (Figure 7).

A recent simulation study under ~ 0 mV membrane potential with a closed gate Na_VAb structure suggested that Na^+ in- and outward movement involves variable configurations of multiple glutamic acid side chains giving rise to non-simple degenerated ion binding modes [23]. Remarkably, detailed investigations of the structural dynamics of E53 in our study revealed distinct isomerization patterns between forward and backward translocations respectively. When the ion moved into the SF from the outer vestibule under hyperpolarized membrane potential, the E53 remained mostly in the non-flipped conformation. In contrast, during outward conduction, the flipping occurrence increased significantly with a typical "one-flip" configuration (Figure 5C and D) when coordinating ions occupied the SF (Figure 7). In our simulations, the depolarized and hyperpolarized membrane potentials of approximately ΔV: 565 mV enabled the detailed investigation of ion permeation directionalities. This was not possible in previous simulations at ~ 0 mV [23]. The conductive, open gate structure used in this study may also reduce the repulsive effect which could have been induced by ions present in the cavity in previous simulations with a closed gate [23]. In addition, different forcefields used in these two studies may also play a substantial role for these discrepancies. As reported by Cordomi et al [29], compared to the combination of OPLS-AA protein with Berger lipids parameters, combining Amber99sb protein and Berger lipids gives more accurate free energies of solvation in water and water to cyclohexane transfer with respect to experimental data for glutamic acid side chains. This may explain the reduced flexibility of the glutamic acid side chain dynamics observed in our study. Further, the force field discrepancies might explain the contrasting results for the "no salt" simulations in these two studies. In the study by Chacrabarti et al [23], the E side chains are more favorable to form flipped conformations even in the "no salt" conformation. This is in contrast to our simulations, where flipping events occurred rarely (Figure 5C) in the "no salt" simulations.

While the inward flow exhibited indistinguishable flux rates irrespective of the E53 conformation (Figure 8A), efflux displayed

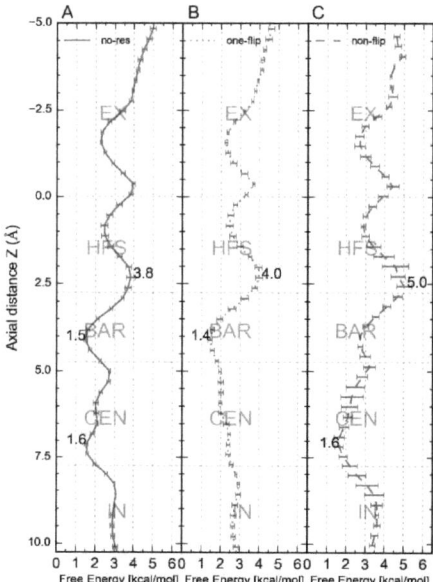

Figure 9. PMF comparison for outward conduction. A) PMF of the simulations without restraints of E53 is depicted as solid line. (B) PMF of the simulations with "one-flip" restraints. C) PMF of the simulations with "non-flip" restraints. The largest free energy barriers in each figure are labeled by terminal peaks and wells. Error estimations shown in the figure are S.E.M.
doi:10.1371/journal.pcbi.1003746.g009

different rates depending on the configurations of the E53 side chain. The highest efflux rate was observed in our "one-flip" simulations and the lowest rate with all four glutamic acid side chains restrained to an outward-facing conformation (Figure 8B). PMF calculations further confirmed that the energy barrier for outward conduction increased from 2.3 kcal/mol to 3.4 kcal/mol when the flipping conformation is prohibited (Figure 9C). That indicates that this flipping conformation provides direct coordination for Na$^+$ ions, which lowers the energy barrier and aids outward conduction.

A simulation study published [30] after the submission of this manuscript, indicates that the SF dynamics, especially the side chain conformational changes of the EEEE locus in the SF, may lead to the conformational changes of the cavity lining helix on the μs timescale, subsequently initiating slow inactivation in Na$_V$ channels. We hypothesize that the E53 dynamics under depolarizing potentials uncovered in this study provide further insights into slow inactivation, especially the fast slow inactivation for prokaryotic species during action potentials. When the membrane potential depolarizes, the probability of Na$^+$ outward transitions increases. As a result, the inactivation probability of the channel is increased probably due to a series of conformational changes starting from the EEEE locus in the SF.

A general limitation of current force fields is that the simulated linear current–voltage regime can only be achieved at higher membrane potentials compared to experimental conditions, resulting from the large electrostatic barriers in the transmembrane region [31,32]. It should be noted that the computational electrophysiology simulations in this study were not done at constant membrane potentials (565±126 mV). This may result from the movement of the ions inside the channels and the fluctuation of the ions in the aqueous compartments (Figure 2A and B). However, a single ion permeation event under physiological conditions will also exist as a non-equilibrium process. Thus, to which extent, current simulation methods resemble ion channels' electrophysiology needs to be further validated. In addition, inaccuracies in the interaction parameters (from the forcefield) between ions and surrounding atoms could also influence the conduction rates [33]. Thus, further structural and computational studies (including optimized strategies for ion interaction with surrounding atoms and polarizable force fields with different lipid species) will be required to further investigate the conformational changes of the SF under different electrochemical drives and the influence of different protonation states of the EEEE locus. In addition, experimental validation is essential to further uncover the structural determinants and the importance of the protonation states of the EEEE locus on ion conductance and selectivity.

Summarizing, our simulations, using applied membrane potentials, reveal different conduction mechanisms for ion inward and outward transitions respectively. An inward facing conformation (flip) of one glutamic acid side chain in the SF would reduce the energy barrier for ion outward transition by providing direct coordination with interacting Na^+ ions. This local change can provide insights into the slow inactivation of Na_V channels as suggested by Boiteux et al [30] during an action potential, when the membrane potential is depolarized.

Methods

MD simulations

The coordinates of $Na_V Ms$ (PDB Entry: 4F4L; Resolution: 3.49 Å) [7] with a conductive pore gate were used. The symmetric tetrameric structure consists of residues 8 to 94. All charged residues were treated keeping their charge states at physiological pH 7.4. In order to investigate ion conductance under two opposite membrane potentials, we used the computational electrophysiology method developed by Kutzner et al. [27] with a double-bilayer scheme. Each bilayer leaflet consists of 242 1-palmitoyl-2-oleoyl-sn-glycero-3-phosphocholine (POPC) lipids encompassing the protein structure, solvated with 500 mM NaCl solution. The system was then duplicated in the Z direction (pore axis). The virtual-site model was adopted for hydrogen atoms [34].

MD simulations were performed with Gromacs version 4.5.5-dev [35,36]. Simulations were carried out with the AMBER99sb [37] all atom force field, POPC lipids parameter were taken from Berger et al. [29,38] with the TIP3P water model [39]. All covalent bonds were constrained using the LINCS algorithm [40], allowing for an integration time step of 4 fs with virtual sites. A 10 Å cutoff was adopted for calculating short-range electrostatic interactions and the Particle Mesh Ewald [41] summation was used for calculating long-range electrostatic interactions. The corrected monovalent ion Lennard-Jones parameters for the amber forcefield [42] were implemented in this study and the vdW interactions were calculated with a cutoff of 10 Å. The Nose-Hoover thermostat [43,44] and the semi-isotropic Parrinello-Rahman barostat algorithm [45] was used to maintain simulation temperature and pressure constantly at 300 K and 1 bar, respectively.

Prior to MD simulations, 3000 conjugate gradient energy-minimization steps were performed, followed by 5 ns equilibration in order to fully solvate mobile water and lipids around the restrained protein with a force constant of 1000 kJ/mol/nm² on all heavy atoms. Hereafter, an equal number of Na^+ ions and a net difference of 4 e of Cl^- across each lipid bilayer between the central electrolyte bath and the two outer ones were sustained during the simulation by a swapping mechanism [27]. In this scheme, a new form of Poisson equation [46] was adopted to derive the potential profile as a function of system length (z). The well-defined transmembrane voltage across each lipid patch was directly assessed by twice integration of this sustained charge density differences between the central electrolyte bath and the two outer ones [27]. In our simulations, depolarized and hyperpolarized membrane potentials were calculated as $\Delta V = 565 \pm 126$ mV (Figure 2). Harmonic restraints (1 kcal/mol/Å²) were exerted on the α-carbon atoms of the TM helices (S5 and S6) throughout the simulations to maintain the open configuration in the absence of the voltage-sensing domain as suggested by Ulmschneider et al. [22]. Four times 500 ns MD simulations were performed; the first 100 ns were treated as equilibration. Simulation trajectories were saved every 100 ps; as a result, 4000 snapshots (analyzing windows) were recorded for analyzing data. Three repeated 500 ns simulations (400 ns were adopted for analysis) with only ions neutralizing the system and no ions in the SF ("no salt") were performed as control to investigate the influence of the membrane potential on the E53 side chain dynamics. In this setup, only ions used to generate the charge imbalance and neutralize the net charge were kept. Further, four repeated 300 ns simulation (200 ns were adopted for analysis) for "non-flip" and "one-flip" configurations respectively were carried out, where the dihedral restraints were applied on the χ_2 dihedral of E53 for all four subunits with a force constant of 500 kcal/mol/rad². This allows dynamic ranges of $56 \pm 10°$ for "one-flip" configuration and $288 \pm 10°$ for "non-flip" configuration of the χ_2 angle.

Ionic binding modes

The total number of ions (i) which completed their conduction in the SF were analyzed (i = 158, from four inward simulations; and i = 79 from four outward simulations). All snapshots of the probe ions (yellow) and coupling ions in the SF (blue) were rendered for five different interaction sites (S_{EX}, S_{HFS}, S_{BAR}, S_{CEN} and S_{IN}). For sites S_{HFS}, S_{BAR} and S_{CEN}, the side chain isomerization states were separated into flipped and non-flipped categories and analyzed. For flipped ones, all four protein chains and the relative ion positions were aligned to chain A to achieve a better representation of the ionic binding patterns. Two probability parameters Pi and Pt were calculated to characterize the influence of E53 dynamics on ion conduction for these three sites. Pi = (Fi/i)*100%, where Fi denotes the number of ions which generated at least one E53 flipping event during their permeation through each site. Pt = Ft/Tt*100%, where Ft denotes the number of snapshots where E53 flipped when the probe ions traversed each site and Tt denotes the total number of snapshots when the probe ions traversed each site.

Potential of mean force (PMF)

The 1-D potential of mean force profile of the ions under membrane potentials were calculated by taking the logarithm of the Na^+ probability distribution along the channel axis (z) in the SF region, according to $G(z) = -k_B T \ln[p(R_i)]$, where k_B is the Boltzmann constant, T is the temperature, and $p(R_i)$ is the probability distribution of the probe ions. 100 bins were used to achieve a bin width of 0.15 Å depicting the details of the profile. Error bars are S.E.M. from four different simulations.

Supporting Information

Figure S1 Cumulative ion flux counts with flip counts as a function of time from simulation 1 without dihedral restrains of E53. A) & C) Ion flux counts through the SF and flipping counts of E53 over 400 ns trajectory of inward simulation (color: blue). B) & D) Ion flux counts through the SF and flipping counts of E53 in over 400 ns trajectory of outward simulation (color: red).
(TIFF)

Figure S2 Cumulative ion flux counts with flip counts as a function of time from simulation 2 without dihedral restrains of E53. A) & C) Ion flux counts through the SF and flipping counts of E53 over 400 ns trajectory of inward simulation (color: blue). B) & D) Ion flux counts through the SF and flipping counts of E53 in over 400 ns trajectory of outward simulation (color: red).
(TIFF)

Figure S3 Cumulative ion flux counts with flip counts as a function of time from simulation 3 without dihedral restrains of E53. A) & C) Ion flux counts through the SF and flipping counts of E53 over 400 ns trajectory of inward simulation (color: blue). B) & D) Ion flux counts through the SF and flipping counts of E53 in over 400 ns trajectory of outward simulation (color: red).
(TIFF)

Figure S4 Cumulative ion flux counts with flip counts as a function of time from simulation 4 without dihedral restrains of E53. A) & C) Ion flux counts through the SF and flipping counts of E53 over 400 ns trajectory of inward simulation (color: blue). B) & D) Ion flux counts through the SF and flipping counts of E53 in over 400 ns trajectory of outward simulation (color: red).
(TIFF)

Figure S5 Hydration shell of the simulations without dihedral restrains of E53. (A) Oxygen coordination numbers in the first hydration shell for inward sodium conduction (Oxygen atoms closer than 3.0 Å to Na^+ were considered coordinating atoms): the total coordination number is depicted in grey; water oxygens are depicted as blue lines. Protein backbone oxygens are shown in green. E53 side chain oxygens are colored red. B) Coordination oxygen atoms numbers for outward sodium conduction.
(TIFF)

Movie S1 The movie shows the double bilayer simulation scheme (400 ns).
(MP4)

Movie S2 A "flipped" trajectory clip showing one example conduction event from an outward simulation. Two opposite subunits are shown as cyan sticks for clarity. The E53 side chains are highlighted in yellow. Sodium ions are shown in yellow, except for Na^+ ions in the SF, which are colored in different colors. The water molecules within 3 Å of these ions are represented as sticks to depict the hydration shell.
(MP4)

Acknowledgments

We thank Tobias Linder and Eva-Maria Zangerl for helpful discussions and critical reading of the manuscript. The computational results presented have been achieved using the Vienna Scientific Cluster (VSC).

Author Contributions

Conceived and designed the experiments: SK ASW. Performed the experiments: SK. Analyzed the data: SK ENT ASW. Contributed reagents/materials/analysis tools: ENT. Wrote the paper: SK ASW.

References

1. Hille B (2001) Ion channels of excitable membranes. Sinauer. 814 p.
2. George AL (2005) Inherited disorders of voltage-gated sodium channels. J Clin Invest 115: 1990–1999. doi:10.1172/JCI25505.
3. Catterall WA (2010) Ion Channel Voltage Sensors: Structure, Function, and Pathophysiology. Neuron 67: 915–928. doi:10.1016/j.neuron.2010.08.021.
4. Payandeh J, Scheuer T, Zheng N, Catterall WA (2011) The crystal structure of a voltage-gated sodium channel. Nature 475: 353–358. doi:10.1038/nature10238.
5. Zhang X, Ren W, DeCaen P, Yan C, Tao X, et al. (2012) Crystal structure of an orthologue of the NaChBac voltage-gated sodium channel. Nature 486: 130–134. doi:10.1038/nature11054.
6. Payandeh J, El-Din TMG, Scheuer T, Zheng N, Catterall WA (2012) Crystal structure of a voltage-gated sodium channel in two potentially inactivated states. Nature 486: 135–139. doi:10.1038/nature11077.
7. McCusker EC, Bagnéris C, Naylor CE, Cole AR, D'Avanzo N, et al. (2012) Voltage of a bacterial voltage-gated sodium channel pore reveals mechanisms of opening and closing. Nat Commun 3: 1102. doi:10.1038/ncomms2077.
8. Bagnéris C, DeCaen PG, Hall BA, Naylor CE, Clapham DE, et al. (2013) Role of the C-terminal domain in the structure and function of tetrameric sodium channels. Nat Commun 4: Available: http://www.nature.com/ncomms/2013/130919/ncomms3465/full/ncomms3465.html. Accessed 18 December 2013.
9. Shaya D, Findeisen F, Abderemane-Ali F, Arrigoni C, Wong S, et al. (2014) Structure of a Prokaryotic Sodium Channel Pore Reveals Essential Gating Elements and an Outer Ion Binding Site Common to Eukaryotic Channels. J Mol Biol 426: 467–483. doi:10.1016/j.jmb.2013.10.010.
10. Tang L, Gamal El-Din TM, Payandeh J, Martinez GQ, Heard TM, et al. (2014) Structural basis for Ca2+ selectivity of a voltage-gated calcium channel. Nature 505: 56–61. doi:10.1038/nature12775.
11. Yue L, Navarro B, Ren D, Ramos A, Clapham DE (2002) The cation selectivity filter of the bacterial sodium channel, NaChBac. J Gen Physiol 120: 845–853.
12. Eisenman G, Horn R (1983) Ionic selectivity revisited: The role of kinetic and equilibrium processes in ion permeation through channels. J Membr Biol 76: 197–225. doi:10.1007/BF01870364.
13. Carnevale V, Treptow W, Klein ML (2011) Sodium Ion Binding Sites and Hydration in the Lumen of a Bacterial Ion Channel from Molecular Dynamics Simulations. J Phys Chem Lett 2: 2504–2508. doi:10.1021/jz2011379.
14. Corry B, Thomas M (2011) Mechanism of Ion Permeation and Selectivity in a Voltage Gated Sodium Channel. J Am Chem Soc 134: 1840–1846. doi:10.1021/ja210020h.
15. Furini S, Domene C (2012) On Conduction in a Bacterial Sodium Channel. PLoS Comput Biol 8: e1002476. doi:10.1371/journal.pcbi.1002476.
16. Qiu H, Shen R, Guo W (2012) Ion solvation and structural stability in a sodium channel investigated by molecular dynamics calculations. Biochim Biophys Acta BBA - Biomembr 1818: 2529–2535. doi:10.1016/j.bbamem.2012.06.003.
17. Ke S, Zangerl E-M, Stary-Weinzinger A (2013) Distinct interactions of Na+ and Ca2+ ions with the selectivity filter of the bacterial sodium channel NaVAb. Biochem Biophys Res Commun 430: 1272–1276. doi:10.1016/j.bbrc.2012.12.055.
18. Zhang X, Xia M, Li Y, Liu H, Jiang X, et al. (2013) Analysis of the selectivity filter of the voltage-gated sodium channel NavRh. Cell Res 23: 409–422. doi:10.1038/cr.2012.173.
19. Xia M, Liu H, Li Y, Yan N, Gong H (2013) The Mechanism of Na+/K+ Selectivity in Mammalian Voltage-Gated Sodium Channels Based on Molecular Dynamics Simulation. Biophys J 104: 2401–2409. doi:10.1016/j.bpj.2013.04.035.
20. Corry B (2013) Na(+)/Ca(2+) selectivity in the bacterial voltage-gated sodium channel NavAb. PeerJ 1: e16. doi:10.7717/peerj.16.
21. Stock L, Delemotte L, Carnevale V, Treptow W, Klein ML (2013) Conduction in a Biological Sodium Selective Channel. J Phys Chem B 117: 3782–3789. doi:10.1021/jp401403b.
22. Ulmschneider MB, Bagnéris C, McCusker EC, DeCaen PG, Delling M, et al. (2013) Molecular dynamics of ion transport through the open conformation of a bacterial voltage-gated sodium channel. Proc Natl Acad Sci: 201214667. doi:10.1073/pnas.1214667110. [epub ahead of print]
23. Chakrabarti N, Ing C, Payandeh J, Zheng N, Catterall WA, et al. (2013) Catalysis of Na+ permeation in the bacterial sodium channel NaVAb. Proc Natl Acad Sci: 201309452. doi:10.1073/pnas.1309452110. [epub ahead of print]
24. Furini S, Domene C (2013) K+ and Na+ Conduction in Selective and Nonselective Ion Channels Via Molecular Dynamics Simulations. Biophys J 105: 1737–1745. doi:10.1016/j.bpj.2013.08.049.
25. Bernèche S, Roux B (2005) A Gate in the Selectivity Filter of Potassium Channels. Structure 13: 591–600. doi:10.1016/j.str.2004.12.019.
26. Cuello LG, Jogini V, Cortes DM, Perozo E (2010) Structural mechanism of C-type inactivation in K+ channels. Nature 466: 203–208.
27. Kutzner C, Grubmüller H, de Groot BL, Zachariae U (2011) Computational Electrophysiology: The Molecular Dynamics of Ion Channel Permeation and Selectivity in Atomistic Detail. Biophys J 101: 809–817.
28. Amaral C, Carnevale V, Klein ML, Treptow W (2012) Exploring conformational states of the bacterial voltage-gated sodium channel NavAb via molecular dynamics simulations. Proc Natl Acad Sci 109: 21336–21341. doi:10.1073/pnas.1218087109.

29. Cordomí A, Caltabiano G, Pardo L (2012) Membrane Protein Simulations Using AMBER Force Field and Berger Lipid Parameters. J Chem Theory Comput 8: 948–958. doi:10.1021/ct200491c.
30. Boiteux C, Vorobyov I, Allen TW (2014) Ion conduction and conformational flexibility of a bacterial voltage-gated sodium channel. Proc Natl Acad Sci 111: 3454–3459. doi:10.1073/pnas.1320907111.
31. Bockmann RA, Hac A, Heimburg T, Grubmuller H (2003) Effect of Sodium Chloride on a Lipid Bilayer. Biophys J 85: 1647–1655.
32. Jensen MØ, Jogini V, Eastwood MP, Shaw DE (2013) Atomic-level simulation of current-voltage relationships in single-file ion channels. J Gen Physiol 141: 619–632. doi:10.1085/jgp.201210820.
33. Luo Y, Roux B (2010) Simulation of Osmotic Pressure in Concentrated Aqueous Salt Solutions. J Phys Chem Lett 1: 183–189. doi:10.1021/jz900079w.
34. Feenstra KA, Hess B, Berendsen HJC (1999) Improving efficiency of large time-scale molecular dynamics simulations of hydrogen-rich systems. J Comput Chem 20: 786–798. doi:10.1002/(SICI)1096-987X(199906)20:8<786::AID-JCC5>3.0.CO;2-B.
35. Hess B, Kutzner C, Van Der Spoel D, Lindahl E (2008) GROMACS 4: Algorithms for highly efficient, load-balanced, and scalable molecular simulation. J Chem Theory Comput 4: 435–447.
36. Van Der Spoel D, Lindahl E, Hess B, Groenhof G, Mark AE, et al. (2005) GROMACS: fast, flexible, and free. J Comput Chem 26: 1701.
37. Hornak V, Abel R, Okur A, Strockbine B, Roitberg A, et al. (2006) Comparison of multiple Amber force fields and development of improved protein backbone parameters. Proteins 65: 712–725. doi:10.1002/prot.21123.
38. Berger O, Edholm O, Jahnig F (1997) Molecular dynamics simulations of a fluid bilayer of dipalmitoylphosphatidylcholine at full hydration, constant pressure, and constant temperature. Biophys J 72: 2002–2013.
39. Jorgensen WL, Chandrasekhar J, Madura JD, Impey RW, Klein ML (1983) Comparison of simple potential functions for simulating liquid water. J Chem Phys 79: 926–935. doi:10.1063/1.445869.
40. Hess B, Bekker H, Berendsen HJC, Fraaije JGEM (1997) LINCS: A linear constraint solver for molecular simulations. J Comput Chem 18: 1463–1472. doi:10.1002/(SICI)1096-987X(199709)18:12<1463::AID-JCC4>3.0.CO;2-H.
41. Darden T, York D, Pedersen L (1993) Particle mesh Ewald: An N log (N) method for Ewald sums in large systems. J Chem Phys 98: 10089–10089.
42. Joung IS, Cheatham III TE (2008) Determination of alkali and halide monovalent ion parameters for use in explicitly solvated biomolecular simulations. J Phys Chem B 112: 9020–9041.
43. Nose S, Nose S (1984) A unified formulation of the constant temperature molecular-dynamics methods. J Chem Phys 81: 511–519. doi:10.1063/1.447334.
44. Hoover, William G., Hoover WG (1985) Canonical dynamics: Equilibrium phase-space distributions. Phys Rev A 31: 1695–1697. doi:10.1103/PhysRevA.31.1695.
45. Martoňák R, Laio A, Parrinello M (2003) Predicting crystal structures: The Parrinello-Rahman method revisited. Phys Rev Lett 90: 75503.
46. Sachs JN, Crozier PS, Woolf TB (2004) Atomistic simulations of biologically realistic transmembrane potential gradients. J Chem Phys 121: 10847–10851. doi:10.1063/1.1826056.

5.4 Paper #4

Probing the Energy Landscape of Activation Gating of the Bacterial Potassium Channel KcsA

Tobias Linder[1], Bert L. de Groot[2], Anna Stary-Weinzinger[1]*

1 Department of Pharmacology and Toxicology, University of Vienna, Vienna, Austria, 2 Computational Biomolecular Dynamics Group, Max Planck Institute for Biophysical Chemistry, Göttingen, Germany

Abstract

The bacterial potassium channel KcsA, which has been crystallized in several conformations, offers an ideal model to investigate activation gating of ion channels. In this study, essential dynamics simulations are applied to obtain insights into the transition pathways and the energy profile of KcsA pore gating. In agreement with previous hypotheses, our simulations reveal a two phasic activation gating process. In the first phase, local structural rearrangements in TM2 are observed leading to an intermediate channel conformation, followed by large structural rearrangements leading to full opening of KcsA. Conformational changes of a highly conserved phenylalanine, F114, at the bundle crossing region are crucial for the transition from a closed to an intermediate state. 3.9 μs umbrella sampling calculations reveal that there are two well-defined energy barriers dividing closed, intermediate, and open channel states. In agreement with mutational studies, the closed state was found to be energetically more favorable compared to the open state. Further, the simulations provide new insights into the dynamical coupling effects of F103 between the activation gate and the selectivity filter. Investigations on individual subunits support cooperativity of subunits during activation gating.

Citation: Linder T, de Groot BL, Stary-Weinzinger A (2013) Probing the Energy Landscape of Activation Gating of the Bacterial Potassium Channel KcsA. PLoS Comput Biol 9(5): e1003058. doi:10.1371/journal.pcbi.1003058

Editor: Emad Tajkhorshid, University of Illinois, United States of America

Received November 22, 2012; **Accepted** March 27, 2013; **Published** May 2, 2013

Copyright: © 2013 Linder et al. This is an open-access article distributed under the terms of the Creative Commons Attribution License, which permits unrestricted use, distribution, and reproduction in any medium, provided the original author and source are credited.

Funding: This work was supported by the Austrian Science Fund (FWF; Grants P22395, W1232; http://www.fwf.ac.at). Tobias Linder was supported by a research fellowship 2013 from the University of Vienna. The funders had no role in study design, data collection and analysis, decision to publish, or preparation of the manuscript.

Competing Interests: The authors have declared that no competing interests exist.

* E-mail: anna.stary@univie.ac.at

Introduction

K$^+$ channels play a crucial role in a wide variety of physiological and pathophysiological processes including action potential modeling [1], cancer cell proliferation [2], and metabolic pathways mediation [3]. In the last few decades, the understanding of ion channels has increased tremendously. The Hodgkin-Huxley equations [4] provided first insights into the ion flow in nerve cells and Hille showed a comprehensive picture of the electrophysiological properties of ion channels [5]. In 1998, the first crystal structure of an ion channel, the bacterial potassium channel of *Streptomyces lividans* (KcsA), shed light on the molecular details of a K$^+$ channel [6]. The pore-forming domain of KcsA is composed of four identical subunits (SUs) which are arranged symmetrically around a channel pore. Each SU consists of two transmembrane helices, TM1 and TM2, which are connected by the P-helix and the selectivity filter (SF) (Figure 1B). While the extracellular facing SF tunes the selection of different ions and modulates inactivation, the main conformational changes regulating ion flow, are found at the TM2 helices. These motions, referred to as activation gating, are thought to involve an iris-like motion of the TM2 helices that constrict the permeation pathway at the helix bundle crossing region [7–10]. This region is believed to form the main activation gate. Starting in 1998, several different pore domain structures of KcsA in its closed state [6,11] and more recently in intermediate and open states have been solved [12]. These crystal structures provide excellent insights into different conformations of proteins; however, they feature only snapshots of dynamical proteins [13].

Therefore, the transition steps and the mechanisms of activation gating are still unknown.

A number of computational studies have been published over the last years, aiming at exploring the gating pathways of ion channels by making use of available X-ray structures as templates [14–22]. However, the lack of particular K$^+$ channels in different conformations was a limitation of previous publications. Thus, these studies had to compare crystal structures of different channels or had to rely on homology models of open structures of KcsA. With the successful crystallization of intermediate and open structures of KcsA by Cuello et al. in 2010 [12], *in silico* activation gating of K$^+$ channels cannot only be readdressed, but also allowed us to calculate a complete energy profile of activation gating. The essential dynamics (ED) simulation method has been shown as a useful tool to investigate sampling of proteins in conformational space and to derive transition pathways between conformational states [23–27]. In this study, we applied ED simulations combined with umbrella sampling calculations to investigate activation gating of KcsA.

Results/Discussion

Stability of closed and open conformations

A prerequisite of the ED method is that the starting and target structures are of equal length and identical amino acid sequence. Thus, the KcsA crystal structures (pdb identifier: 1k4c, closed; 3f6b, intermediate; 3f7v, open) were adjusted at the N- and C-termini so that all states started from residue 29 and ended at

Author Summary

Voltage gated ion channels are membrane embedded proteins that initiate electrical signaling upon changes in membrane potential. These channels are involved in biological key processes such as generation and propagation of nerve impulses. Mutations may lead to serious diseases such as cardiac arrhythmia, diabetes or migraines, rendering them important drug targets. The activity of ion channels is controlled by dynamic conformational changes that regulate ion flow through a central pore. This process, which involves opening and closing of the channels, is known as gating. To fully understand or to control ion channel gating, we need to unravel the underlying principles. Crystal structures, especially of K⁺ channels, have provided excellent insights into the conformation of different channel states. However, the transition states and structural rearrangements are still unknown. Here we use molecular dynamics simulations to simulate the full transition pathway and energy landscape of gating. Our results suggest that channel gating involves local structural changes followed by global conformational changes. The importance of many of the residues identified in our simulations is supported by experimental studies. The ability to accurately simulate the gating transitions of ion channels may be beneficial for a better understanding of ion channel related diseases and drug development.

---

residue 118, leading to channels with four times 89 amino acids. Additionally, Q117 in the open and intermediate crystal structure was mutated to arginine to obtain the wild type structure.

Before probing the transition pathway between closed and open conformations of KcsA, the stability of the different channel states was assessed in molecular dynamics (MD) simulations. Repeated simulations (3 times 50 ns) of the structures, embedded in a lipid-bilayer membrane, were performed. The root-mean-square deviation (RMSD) of the backbone atoms without loops of all three channel states is less than 2 Å (Figure S1). The stability of the closed state is similar to previous values reported in literature [28,29]. Moreover, the RMSD of the intermediate state is comparable to the two other states with a RMSD of 1.75 Å.

Activation gating simulated by essential dynamics

To investigate the activation pathway, the backbone atoms of closed and open structures without loops were compared by principal component analysis (PCA). The resulting eigenvector (EV) was used to enforce the transition between the two states. Thus, the ED simulation is a free MD simulation, with all coordinates equilibrating except for one coordinate that is biased to drive the gating transition. Ten opening and ten closing ED simulations, all of them lasting for 20 ns, were carried out. In the following paragraphs, results of opening simulations are explained in detail. Since similar observations were also found in the reversed direction, results for the closing runs are summarized at the end of this section and corresponding figures are shown in the supplemental material.

The conformational changes during the ED opening simulations were analyzed by monitoring the RMSD as a function of time (Figure 1A). The deviation from the target structure (open conformation, pdb identifier: 3f7v) was measured over time. The difference between the starting and target structure is 4 Å. In all ten opening ED simulations, the RMSD values steadily decreased and reached final values between 1.35 and 2.20 Å, indicating that all simulations reached the open state. Successful opening is defined by a decrease of the RMSD to approximately 2 Å compared to the target structure. For simplicity, the average RMSD and standard deviation of the ten simulations were calculated. On average, a final RMSD of 2 Å as shown in Figure 1A was reached. The standard deviation indicates that in the first 11 ns, the RMSD values of the simulations did not vary. However, in the subsequent simulation time at which the simulations reached the target structure, the RMSD of the ten simulations showed wider distribution.

To investigate the conformational states of the end structures, the deviation of the Cα atoms from the target structure was analyzed. An average structure of the ten ED simulations was generated which exhibits minimal RMSD (Figure 1B). This average structure revealed that ED simulations were able to reach the target structure. Figure 1B shows the color coded deviation of each Cα atom from the open structure. As expected, the TM1 and P-helices displayed a very modest RMSD deviation of 0.05 Å to the target structure since there are no conformational changes in these regions during activation gating. In contrast, deviations up to 3 Å were found in the C-termini of the TM2 helices, which undergo large conformational changes during channel opening. Additionally, large deviations were found in the loop regions due to the high mobility of loops. Investigations on the loop region (amino acid G56) showed that mutations did not influence gating [30,31]. Thus, the loops were not investigated further.

The program HOLE [32] was used to calculate the activation gate radius profiles (Figure 2) of the backbone atoms of different channel states. In the closed conformation, the constriction of the activation gate features a diameter of 5.9 Å. In the intermediate state, the diameter of the constriction site is 8.3 Å. In the open conformation, the activation gate diameter expands to 11.8 Å. The diameter of the activation gate in the ED simulations reached 10.7 Å on average. The shape of the pore radius profile of the end structures obtained from ED simulations matched the essential features of the profile of the open crystal conformation, further indicating that the simulation derived structures adopted the open state.

The major motions of opening were also observed in the reversed direction during closing (see Figure S2). However, only seven out of ten ED simulations successfully closed (RMSD<2.3 Å). Careful inspection revealed that the underlying reason for unsuccessful closure of three runs was partial unwinding of single TM2 helices. This observation may suggest that optimal packing of helices at the bundle crossing region is important for channel closure.

Coupling between activation gate and SF

As described in the method section, no forces were applied to the side chains in the simulations. Hence, the simulations allowed investigations of the rotameric side chain changes coupled to gating. A phenylalanine, F103, present in the TM2 helices of KcsA, was shown to change its rotameric state upon activation gating [12,21] and affecting the SF conformation [33,34]. Therefore, the χ_1 angle dynamics in the ten ED simulations were analyzed (Figure 3A). F103 can adopt two different rotameric states which are called "up" (χ_1 angle of -55 to $-72°$) and "down" state (χ_1 angle of -166 to $-185°$). In the first 5 ns of the opening ED simulations, F103 was stable in the up state. Subsequently, the conformational changes of the channel allowed F103 to adopt the down state. The F103 amino acids switched from the up to the down state over the next 15 ns. In most of the cases, this change was irreversible. Once F103 was in the down state, it was not able to switch to the up state again. After 20 ns, 78% of all F103 were in the down state. To validate if the F103 rotameric changes occurred because of activation gating, dihedral

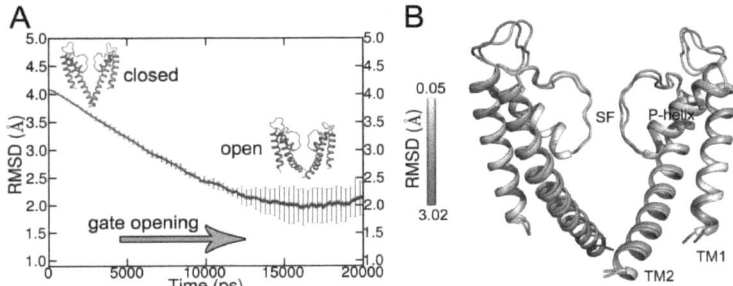

Figure 1. RMSD analysis of ED opening simulations. A) Average of the backbone RMSD (without loops) of ten opening ED simulations. The open crystal structure was used as reference. The standard deviation is indicated by error bars. B) Comparison of the average structure (built out of the minimal RMSD structures of the ten ED simulations; yellow to red) and the open crystal structure (gray). The RMSD of the Cα atoms is shown as a spectrum from yellow to red. For the sake of clarity, only the two opposite SUs are shown.
doi:10.1371/journal.pcbi.1003058.g001

angles of unbiased open and closed state MD simulations were analyzed (data not shown). In the open state, all F103 of the three 50 ns MD simulations were in the down state. In the closed conformation, F103 showed more flexibility. Initially in the up state, the F103 was able to change to the down state; however, the up state is observed more frequently. This finding is in agreement with adiabatic energy maps of Pan et al [34] and a study by Cuello et al [33]. The dynamic behavior of F103 in the closing ED simulations is shown in Figure S2. In the first 2 ns, F103 was stable in the down state. Subsequently, F103 can adopt both up and down states as expected from the energy maps of Pan et al [34].

Despite different SF conformations in the closed and open crystal structures (actived vs. inactivated), the SF in all ten opening simulations did not adopt the inactivated conformation as seen in the crystal structure (pdb identifier: 3f7v; Figure 3B–E). The stability of the SF of the ED derived open conformation is further supported by a 100 ns free MD simulation, where no changes in the filter were observed. Previous studies reported that side chain

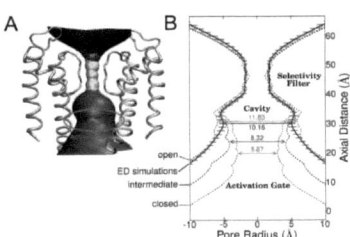

Figure 2. Pore radius profiles derived from backbone atoms of channel states. A) 3D representation of the pore domain depicting the HOLE profile. For the sake of clarity, only two opposing SUs are shown. B) Comparison of the profiles formed by the closed (red dashed line), intermediate (blue dashed line), and open (green dashed line) crystal structures with the average of the ten ED simulation structures (black dashed line). The subtle differences of the ED simulation structures in the activation gate region are indicated as standard deviation by error bars.
doi:10.1371/journal.pcbi.1003058.g002

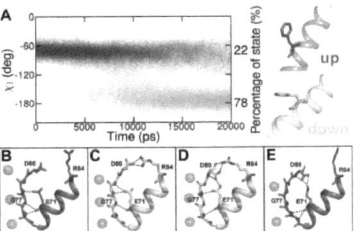

Figure 3. Conformational changes of F103 during activation gate opening and SF conformations of channel states. A) Analysis of χ_1 angle dynamics of F103 (in ten opening ED simulations). Changes of the F103 orientation (χ_1 angle) were measured over time. An angle of $-70°$ indicates the "up" state (blue) while an angle of $-180°$ represents the "down" state (yellow). The percentage of state was calculated from the end states at 20 ns of the ten ED simulations. B) SF and P-helix of the closed crystal structure (gray). Blue spheres represent K⁺ ions. C) SF and P-helix at the end of the 20 ns ED simulation structure with deprotonated E71 (yellow). D) SF and P-helix at the end of the 20 ns ED simulation structure with protonated E71 (yellow). E) SF and P-helix of the open inactivated crystal structure (gray). The G77 conformation defines the SF state as it was shown by Cuello et al. [12].
doi:10.1371/journal.pcbi.1003058.g003

hydrogen bonds between D80 and a protonated E71 promote inactivation of the SF [35–37]. Hence, we performed ED simulations with protonated E71 amino acids and analyzed the SF conformation. These simulations revealed similar conformations, irrespective of the protonation state. This conformation might be influenced by the ion occupancy in the filter. The ions were located at the most favored positions S0, S2, and S4 since the simulations started from a conductive state [38].

Free energy profile of activation gating

Umbrella sampling was employed to investigate the free energy landscape of activation gating (Figure 4A). The ED simulation with the lowest RMSD was used for a subsequent PCA calculation and thereof the first EV was employed as reaction coordinate. MD simulations of closed, intermediate, and open states were projected onto this reaction coordinate to determine sampling regions of the crystal structures. Three main energy wells, separated by two energy barriers, were identified. The first energy well, which is sampled by the closed state, is located at −0.7 to 3.1 nm. The intermediate state is sampled at the adjacent energy well, separated by a small energy barrier at 4 nm (barrier 1) from the closed state. Broad sampling of the intermediate conformation was observed, ranging from 3.4 to 7.4 nm. The subsequent large energy barrier at 9 nm (barrier 2) separates the open conformation from the intermediate state. The open conformation samples a relatively small energy well ranging from 8.8 to 11.4 nm. Next, we investigated the underlying structural rearrangements shaping the energy wells and barriers.

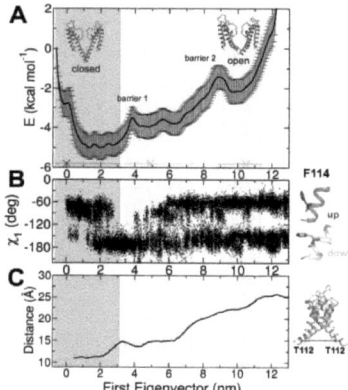

Figure 4. Free energy profile of gate opening. The blue, green, and yellow shades depict the sampling of the closed, intermediate, and open structures along the first EV. A) Free energy profile of the activation gate opening derived from 3.9 μs umbrella sampling simulations. The X marks indicate the positions of the crystal structures. B) χ₁ angle dynamics of all four F114 during activation gating. C) Distance between opposite T112 as a measure of pore opening.
doi:10.1371/journal.pcbi.1003058.g004

Local structural rearrangements correspond to energy barrier 1

By analyzing the dihedral angles of all side chains, a single residue in the helix bundle crossing region was identified (F114) whose conformational changes correspond to the first energy barrier (Figure 4B). This unique rotameric pattern of F114 was observed in all ten opening ED simulation runs suggesting that this pattern was essential for activation gating (Figure 5A). In the early stage of activation gating (after 5 ns), 80% of all F114 changed from an up state (χ_1 angle of −55 to −72°) to a down state (χ_1 angle of −166 to −185°). After the change to the down state, a rigid phase from 5 to 10 ns was observed. Subsequently, F114 regained its flexibility. This suggests that the first flip of F114 and the changes in interacting amino acids may cause energy barrier 1. Consequently, interacting amino acids were analyzed in more detail. Figure 5B–E depicts residues that interact with F114 over time. Residues L110, W113, and R117 of TM2 and L105 of the adjacent TM2 helix interacting in all states are shown in green. Additional interacting amino acids in the closed state were A108, A109, and T112 of the adjacent TM2 (Figure 5B). In the rigid transition state (Figure 5C), additional interactions to V115 were observed. In the open state, interactions with T101 and S102 of the neighboring TM2 were found. When F114 occupied the down state, it was in close contact with A32 of the adjacent TM1 helix. F114 interacted with L35 (adjacent TM1 helix) independently of the rotameric state, indicating a specific interaction pattern. The importance of the F114 and adjacent amino acids is supported by experimental mutation studies (see section "relation to experimental data").

The dynamical behavior of the F114 side chain is further supported by free MD simulations of the open and closed state. In the open state, 75% of the 12 F114 side chains in the MD simulations adopted the down state. Flipping between the two states occurred as a rare event, indicating that the F114 side chains showed high stability over 50 ns. An increased flexibility of F114 was observed in the closed state. Although 80% of the F114 side chains adopted the initial up state, flipping between the two states was observed frequently. Nevertheless, the specific rotameric pattern of F114 as seen during the ED simulations did not occur, indicating that this rotameric pattern is unique for activation gating. Additionally, these analyses showed that not only F103 but also F114 is allowed to adopt two rotameric states in the closed conformation.

Global conformational changes of TM2 correspond to energy barrier 2

Cα-Cα distances between two opposite T112 residues (TM2) as a measure of activation gate opening (as proposed by Cuello et al [12]) were found to correlate with the energy barriers (Figure 4C). This measurement allows direct comparison of ED derived conformational states (closed, intermediate, and open) to the crystal structures. At the first energy barrier, an initial conformational change of the activation gate from 12 Å to 14 Å was observed correlating to structural rearrangements of F114. In the subsequent plateau phase of opening, a good correlation with the energy wells of the intermediate structures was observed. The second energy barrier is linked to a distance increase of 8 Å between the two opposing T112 residues. This suggests that the second energy barrier is mainly caused by global conformational changes of TM2. To further test the significance of this two-phasic activation gate opening, the T112 distances of all ten opening ED simulations were analyzed. Again, a two-phasic gating with global conformational changes at 4 to 5 ns and at 7.5 to 16 ns was found

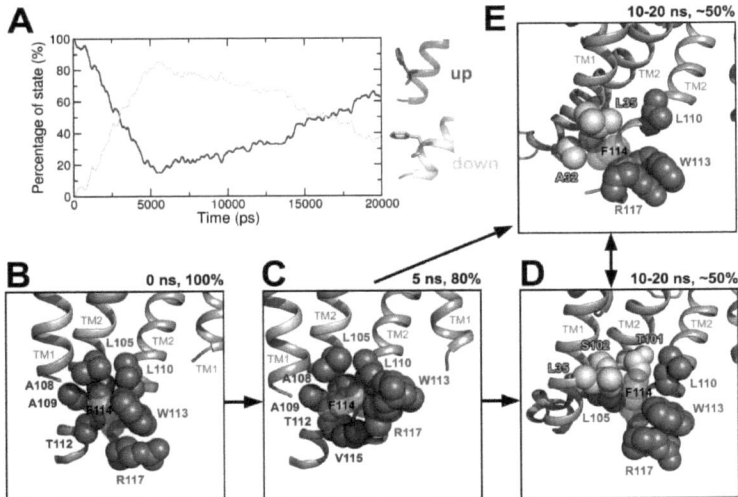

Figure 5. Analysis of χ_1 angle dynamics of F114 and influence on packing. A) Percentage of F114 in the up (blue) and down (yellow) state over time. Packing of the F114 (transparent orange spheres) in the closed conformation (B), the transition state (C), the open states with F114 in the up (D) and the down state (E). Amino acids interacting in all states are shown in green. Interacting amino acids in the closed/open/transition state are represented in blue/yellow/red.
doi:10.1371/journal.pcbi.1003058.g005

(Figure S3). These findings are in line with previous computational studies, which showed that the main opening of the gate occurs after an initial unlock from the closed state by structural rearrangements of amino acids [18,21]. Additionally, simulations in the reverse direction showed similar local and global structural rearrangements in inverse order supporting the validity of the simulations.

Relation to experimental data

The transition pathways obtained by the ED simulations are in good agreement with experimental data. First, the simulations are able to sample the intermediate crystal structure (pdb identifier: 3f7v; green shaded energy well in Figure 4A) [12], which was not included in our ED simulation protocol. Secondly, as expected [13], KcsA crystal structures 1k4c, 3f7v, and 3f b6 occupy energy wells in our calculated energy profile (Figure 4A). Thirdly, the energy profile indicates that the pore is intrinsically more stable in the closed conformation. This observation is supported by experimental studies on potassium channels [39–41], although it should be noted that the latter two studies were carried out on shaker-like channels, rendering the comparison indirect. Further, residues involved in pH sensing of KcsA were not included in the simulated system, which may also affect stability.

Simulations support the hypothesis that the F114 conformational changes are crucial to trigger initial activation gating.

Mutational studies have shown the important role of the tightly packed helix bundle crossing region including F114. Several mutations in this region revealed a destabilization of the closed conformation [39,42]. The fact that F114 is conserved in many K$^+$ channels additionally underlines the importance of this aromatic amino acid for channel function [40,43–46]. Mutational analysis of interacting amino acids in the open state like L35, T101, and T102 (analyzed in Shaker [40,47]) or A32 would be of great interest and may lead to new insights into the packing of F114 in the open state.

Lipid interactions of TM2 helices during activation gating

Since the C-terminus of the TM2 helices moves from a water environment towards the lipid/water interface during activation gating, interactions between the TM2 helices and lipids were investigated. Analyses revealed that the number of hydrogen bonds between the hydrogen bond forming residues W113 and R117 and the lipid head groups increased during gate opening (Figure 6). This indicates that the C-terminus of TM2 moved towards the inner leaflet of the bilayer membrane while hydrogen bonds are mainly formed between R117 and the phosphate groups of the lipids. A decrease of hydrogen bonds was found for the closing simulations (Figure S4) while TM2 moves back from the lipid environment to the water environment.

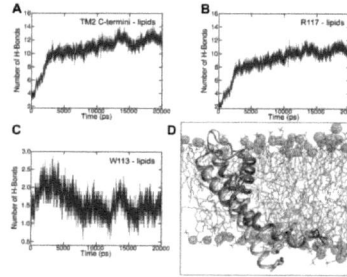

Figure 6. Lipid interactions of TM2 helices during activation gate opening. A) Average number of H-bonds between H-bond forming residues (W113 and R117) of the C-terminal TM2 helices and lipid head groups was measured over time. B) Average number of H-bonds of R117 with lipids. C) Average number of H-bonds of W113 with lipids. D) Representation of one SU in the closed (light blue) and open (marine blue) conformation with lipids. H-bond forming residues W113 and R117 are shown as yellow sticks. Lipids are depicted as gray lines while phosphate groups are shown as orange spheres. Dashed black lines represent H-bonds.
doi:10.1371/journal.pcbi.1003058.g006

Cooperativity of activation gating

ED simulations were applied on one, two, and three SUs, respectively, while the other SUs were allowed to move freely. Simulations revealed that at least three SUs are necessary to open the activation gate. RMSD analyses of simulations with the ED method applied on one and two SUs showed that there was only a slight decrease in RMSD over time suggesting that the channel remained in the closed state. However, simulations with the ED method applied on three SUs revealed that the end structures deviated 2.5 Å from the target structure (Figure 7). Cooperativity analyses of ED simulations presented in this study support previous studies on cooperativity of potassium channels in general [48–51] and of the pore domain in particular [19,21,52,53]. Our simulations indicate that movement of one SU or two SUs is insufficient to open the gate. However, opening of three SUs is sufficient to obtain an open gate structure. Comprehensive investigations on cooperativity are subject of further studies.

Conclusion

The results presented here show that the ED simulation approach successfully sampled transition pathways between closed and open states of an ion channel on the nanosecond time scale and allowed investigations on activation gating. There is good agreement between our investigations and previous experimental and computational studies, supporting the validity of this approach. The simulations provided new insights into conformational changes during gating and revealed that activation gating occurs as a two phase process. Additionally, investigation of the energy landscape allowed the correlation of conformational changes to energy barriers at the atomistic level. The first phase, in which local structural rearrangements in the helix bundle crossing region take place, correlates to a small energy barrier. The second phase was found to correlate with a large second energy barrier. During this phase, the main conformational changes of the TM2 helices, which occur upon gating, were observed.

In addition, we showed the feasibility of the ED approach to study the cooperativity of activation gating. The simulations suggest that individual SUs cannot open the activation gate. Rather, several SUs have to move in a cooperative manner in order to open the gate.

We expect that ED simulations will be useful for further investigations including the analysis of gating sensitive mutations. This is of special interest with regard to inherited channelopathies. Furthermore, we expect that these simulations will be valuable for studies on drug binding with different channel states.

Methods

Simulation setup

The closed (pdb identifier: 1k4c) [54] and open (pdb identifier: 3f7v) [12] crystal structures were used as starting conformations for the ED simulations. Additionally, they were subject to free MD simulations to assess the stability and the side chain dynamics. Free MD simulations of the intermediate conformation (pdb identifier: 3f5b) [12] were performed to investigate the sampling region of the structure along the transition pathway. Since the helices of the open conformation were not crystallized to the same extent as in the closed state (seven amino acids are missing at the beginning of TM1 and six amino acids at the end of TM2), the helix-lengths of the closed crystal structure were adapted by deleting these amino acids. The Q117 in the crystal structure of the open conformation was mutated to arginine in order to obtain the wild type structure using Swiss-PdbViewer [55]. For the intermediate conformation, one helical turn on the C-terminus was added in PdbViewer to obtain the same length of the helices as for the closed and open conformation. The protein was embedded in an equilibrated membrane consisting of 280 dioleoylphosphatidylcholine (DOPC) lipids using the g_membed tool [56], which is part of the gromacs package. K^+ ions were placed in the SF, as described previously [57], at K^+ sites S0, S2, and S4, with waters placed at S1 and S3 of the SF [38]. Cl^- ions were added randomly within the solvent to neutralize the system. All simulations were carried out using the gromacs simulation software v.4.5.4 [58]. The amber99sb force field [59] and the TIP3P model [60] were employed for the protein and water, respectively. Lipid parameter for the DOPC membrane were taken from Siu et al [61]. During all simulations, the area per lipid was at 0.72 nm^2 which is in good agreement with experimental values [62]. Electrostatic interactions were calculated at every step with the particle-mesh Ewald method [63] with a short-range electrostatic interaction cut off of 1.4 nm. Lennard-Jones interactions were calculated with a cut off of 1.4 nm. The LINCS algorithm [64] was used to constrain bonds, allowing for an integration step of 2 fs. The Nose-Hoover thermostat was used to keep simulation temperature constant by coupling (tau = 0.5 ps for equilibration simulations and tau = 0.2 ps during unrestrained simulations) the protein, lipids and solvent (water and ions) separately to a temperature bath of 310 K. Likewise, the pressure was kept constant at 1 bar by using the Parrinello-Rahman barostat algorithm with a coupling constant of 1 ps. Prior to simulation, 1000 conjugate gradient energy-minimization steps were performed, followed by 5 ns of equilibrium simulation in which the protein atoms were restrained by a force constant of 1000 kJ mol^{-1} nm^{-2} to their initial position. Lipids, ions, and water were allowed to move freely during equilibration.

Molecular dynamics simulations

In order to assess the stability of the open, intermediate, and closed conformation of the KcsA channel, three 50 ns unrestrained MD simulations were carried out for each structure.

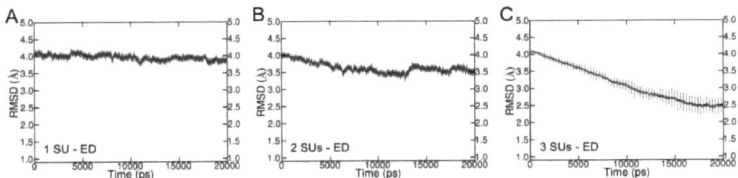

Figure 7. RMSD of cooperativity simulations. ED was applied on one (A), two (B), and three (C) SUs. The open structure was used as reference. For simulations with ED applied on one SU and two SUs, only one simulation each was performed. For ED simulations on three SUs, the average of the backbone RMSD without loops of ten simulations was measured. Standard deviation is indicated by the error bars.
doi:10.1371/journal.pcbi.1003058.g007

Principal component analysis

The basic method of the PCA is described in detail elsewhere [65]. A trajectory consisting of the closed and the open conformation was built and used for PCA. Subsequently, the covariance matrix of the positional fluctuations of the TM1, P-helix, and TM2 backbone atoms was built up and diagonalized (loops were excluded from analysis). For the PCA, all four SUs (one, two, and three SUs for cooperativity investigations) of the homotetrameric channel were taken into account. Only one EV with a non-zero eigenvalue results from this PCA, which represents the difference vector between the open and the closed crystal conformation. This vector was used as reaction coordinate for ED simulations.

Essential dynamics simulations

The ED technique [23,24] can be used to simulate the conformational pathway between two crystal structures [26]. During simulation, the distance along the first EV was increased in fixed increments to drive the system from the closed to the open state and vice versa. It is important to emphasize that the EVs were obtained by PCA of the backbone atoms only and therefore did not contain any information on the side chains. For simulations, the equilibrated closed and open systems, respectively, consisting of the channel, lipid-membrane, ions, and water, were used as start positions. Helical restraints were applied to the last four C-terminal amino acids of the TM2 helix of each SU in order to prevent unwinding. All parameters were set as described above. Simulations were performed on the 20 ns timescale. Fixed increment linear expansion for each simulation step (2 fs) was set to $1.28e^{-6}$ nm in order that the target structure was reached after two thirds of the simulation time. For cooperativity investigations, fixed increment linear expansion was set to $1.89e^{-7}$ nm, $6.27e^{-7}$ nm, $9.24e^{-7}$ nm per step (2 fs) and was applied to one SU, two SUs, and three SUs, respectively.

Umbrella sampling

The windows for the umbrella sampling simulation were taken from the ED simulation with the lowest RMSD. The first EV, which was derived from a PCA of the ED simulation, was used as a reaction coordinate. As this EV is dominant (its eigenvalue is more than an order of magnitude larger than the second largest), we assume that the transition pathway is sufficiently accurately covered by this mode. Along this reaction coordinate, 39 windows with the corresponding structures from the first ED simulation were chosen for umbrella sampling and simulated for 100 ns (Figure S5). 33 windows were simulated with a force constant of 1 kJ mol^{-1} nm^{-2}. For six windows, the force constant was set to 100 kJ mol^{-1} nm^{-2} in order to obtain sufficient sampling of the energy barriers. In total, umbrella sampling was performed for 3.9 µs. The first 50 ns of each window were discarded for equilibration. The potential of mean force and the statistical errors of the activation gating energy profile were estimated by making use of the g_wham tool of gromacs and the integrated bootstrap analysis method [66]. The number of bootstraps was set to 50.

Supporting Information

Figure S1 Stability of KcsA channel states. Backbone RMSD (without loops) of three independent MD simulations of closed (A), intermediate (B), and open state (C) was measured as a function of time.
(TIF)

Figure S2 Analysis of ED closing simulations. A) Average of the backbone RMSD without loops of seven closing ED simulations. The closed crystal structure was used as reference. The standard deviation is indicated by error bars. B) Conformational changes of F103 during activation gate closing. Analysis of χ_1 angle dynamics of F103 of the seven ED simulations was performed. Changes of the F103 orientation was measured as χ_1 angle over time. An angle of $-70°$ indicates the "up" state (blue) while an angle of $-180°$ represents the "down" state (yellow). C) χ_1 angle dynamics of F114 are shown as percentage of F114 in the up (blue) and down (yellow) states over time.
(TIF)

Figure S3 Average of the Cα-Cα T112-distances of all ten ED simulations. The standard deviation is indicated by error bars.
(TIF)

Figure S4 Lipid interactions of TM2 helices during activation gate closing. A) Average number of H-bonds between H-bond forming residues (W113 and R117) of the C-terminal TM2 helices and lipid head groups was measured over time. B) Average number of H-bonds of R117 with lipids. C) Average number of H-bonds of W113 with lipids.
(TIF)

Figure S5 Histograms of the 39 umbrella sampling windows. The six windows with peaks above 40000 were derived from umbrella sampling with a force constant of 100 kJ mol^{-1} nm^{-2} (default: 1 kJ mol^{-1} nm^{-2}).
(TIF)

Acknowledgments

The computational results presented have been achieved in part using the Vienna Scientific Cluster (VSC). The authors would like to thank Song Ke, Eva-Maria Zangerl, and Julia Praxmarer for critical reading of the manuscript.

Author Contributions

Conceived and designed the experiments: TL BLdG ASW. Performed the experiments: TL. Analyzed the data: TL BLdG ASW. Wrote the paper: TL BLdG ASW.

References

1. Pollard CE, Abi Gerges N, Bridgland-Taylor MH, Easter A, Hammond TG, et al. (2010) An introduction to QT interval prolongation and non-clinical approaches to assessing and reducing risk. Br J Pharmacol 159: 12–21. doi:10.1111/j.1476-5381.2009.00207.x.
2. Jehle J, Schweizer PA, Katus HA, Thomas D (2011) Novel roles for hERG K(+) channels in cell proliferation and apoptosis. Cell Death Dis 2: e193. doi:10.1038/cddis.2011.77.
3. Nichols CG (2006) KATP channels as molecular sensors of cellular metabolism. Nature 440: 470–476. doi:10.1038/nature04711.
4. Hodgkin AL, Huxley AF (1952) A quantitative description of membrane current and its application to conduction and excitation in nerve. J Physiol 117: 500–544.
5. Hille B (2001) Ion channels of excitable membranes. 3rd edition. Sunderland: Sinauer.
6. Doyle DA, Morais Cabral J, Pfuetzner RA, Kuo A, Gulbis JM, et al. (1998) The structure of the potassium channel: molecular basis of K+ conduction and selectivity. Science 280: 69–77. doi:10.1126/science.280.5360.69.
7. Perozo E, Cortes DM, Cuello LG (1999) Structural rearrangements underlying K+-channel activation gating. Science 285: 73–78. doi:10.1126/science.285.5424.73.
8. Kelly BL, Gross A (2003) Potassium channel gating observed with site-directed mass tagging. Nat Struct Biol 10: 280–284. doi:10.1038/nsb908.
9. Zimmer J, Doyle D a, Grossmann JG (2006) Structural characterization and pH-induced conformational transition of full-length KcsA. Biophys J 90: 1752–1766. doi:10.1529/biophysj.105.071175.
10. Shimizu H, Iwamoto M, Konno T, Nihei A, Sasaki YC, et al. (2008) Global twisting motion of single molecular KcsA potassium channel upon gating. Cell 132: 67–78. doi:10.1016/j.cell.2007.11.040.
11. Uysal S, Vásquez V, Tereshko V, Esaki K, Fellouse FA, et al. (2009) Crystal structure of full-length KcsA in its closed conformation. Proc Natl Acad Sci U S A 106: 6644–6649. doi:10.1073/pnas.0810663106.
12. Cuello LG, Jogini V, Cortes DM, Perozo E (2010) Structural mechanism of C-type inactivation in K(+) channels. Nature 466: 203–208. doi:10.1038/nature09153.
13. Henzler-Wildman K, Kern D (2007) Dynamic personalities of proteins. Nature 450: 964–972. doi:10.1038/nature06522.
14. Biggin PC, Sansom MSP (2002) Open-state models of a potassium channel. Biophys J 83: 1867–1876. doi:10.1016/S0006-3495(02)73951-9.
15. Tikhonov DB, Zhorov BS (2004) In silico activation of KcsA K+ channel by lateral forces applied to the C-termini of inner helices. Biophys J 87: 1526–1536. doi:10.1529/biophysj.103.037770.
16. Compoint M, Picaud F, Ramseyer C, Girardet C (2005) Targeted molecular dynamics of an open-state KcsA channel. J Chem Phys 122: 134707. doi:10.1063/1.1869413.
17. Shrivastava IH, Bahar I (2006) Common mechanism of pore opening shared by five different potassium channels. Biophys J 90: 3929–3940. doi:10.1529/biophysj.105.080093.
18. Enosh A, Raveh B, Furman-Schueler O, Halperin D, Ben-Tal N (2008) Generation, comparison, and merging of pathways between protein conformations: gating in K-channels. Biophys J 95: 3850–3860. doi:10.1529/biophysj.108.135285.
19. Haliloglu T, Ben-Tal N (2008) Cooperative transition between open and closed conformations in potassium channels. PLoS Comput Biol 4: e1000164. doi:10.1371/journal.pcbi.1000164.
20. Mashl RJ, Jakobsson E (2008) End-point targeted molecular dynamics: large-scale conformational changes in potassium channels. Biophys J 94: 4307–4319. doi:10.1529/biophysj.107.118778.
21. Denning EJ, Woolf TB (2010) Cooperative nature of gating transitions in K(+) channels as seen from dynamic importance sampling calculations. Proteins 78: 1105–1119. doi:10.1002/prot.22632.
22. Miloshevsky G V, Jordan PC (2007) Open-state conformation of the KcsA K+ channel: Monte Carlo normal mode following simulations. Structure 15: 1654–1662. doi:10.1016/j.str.2007.09.022.
23. Amadei A, Linssen AB, De Groot BL, Van Aalten DM, Berendsen HJ (1996) An efficient method for sampling the essential subspace of proteins. J Biomol Struct Dyn 13: 615–625. doi:10.1080/07391102.1996.10508874.
24. De Groot BL, Amadei A, Van Aalten DM, Berendsen HJ (1996) Toward an exhaustive sampling of the configurational spaces of the two forms of the peptide hormone guanylin. J Biomol Struct Dyn 13: 741–751. doi:10.1080/07391102.1996.10508888.
25. De Groot BL, Amadei A, Scheek RM, Van Nuland NA, Berendsen HJ (1996) AID-PROT7>3.0.CO;2-D.
26. Van Aalten DM, Conn DA, De Groot BL, Berendsen HJ, Findlay JB, et al. (1997) Protein dynamics derived from clusters of crystal structures. Biophys J 73: 2891–2896. doi:10.1016/S0006-3495(97)78317-6.
27. Narzi D, Daidone I, Amadei A, Di Nola A (2008) Protein Folding Pathways Revealed by Essential Dynamics Sampling. J Chem Theory Comput 4: 1940–1948. doi:10.1021/ct800157v.
28. Anishkin A, Milac AL, Guy HR (2010) Symmetry-restrained molecular dynamics simulations improve homology models of potassium channels. Proteins 78: 932–949. doi:10.1002/prot.22618.
29. Bernèche S, Roux B (2000) Molecular dynamics of the KcsA K(+) channel in a bilayer membrane. Biophys J 78: 2900–2917. doi:10.1016/S0006-3495(00)76831-7.
30. Iwamoto M, Shimizu H, Inoue F, Konno T, Sasaki YC, et al. (2006) Surface structure and its dynamic rearrangements of the KcsA potassium channel upon gating and tetrabutylammonium blocking. J Biol Chem 281: 28379–28386. doi:10.1074/jbc.M602018200.
31. Iwamoto M, Oiki S (2013) Amphipathic antenna of an inward rectifier K+ channel responds to changes in the inner membrane leaflet. Proc Natl Acad Sci U S A 110: 749–754. doi:10.1073/pnas.1217323110.
32. Smart OS, Neduvelil JG, Wang X, Wallace BA, Sansom MSP (1996) HOLE: A program for the analysis of the pore dimensions of ion channel structural models. J Mol Graph 14: 354–360. doi:10.1016/S0263-7855(97)00009-X.
33. Cuello LG, Jogini V, Cortes DM, Pan AC, Gagnon DG, et al. (2010) Structural basis for the coupling between activation and inactivation gates in K(+) channels. Nature 466: 272–275. doi:10.1038/nature09136.
34. Pan AC, Cuello LG, Perozo E, Roux B (2011) Thermodynamic coupling between activation and inactivation gating in potassium channels revealed by free energy molecular dynamics simulations. J Gen Physiol 138: 571–580. doi:10.1085/jgp.201110670.
35. Cordero-Morales JF, Cuello LG, Zhao Y, Jogini V, Cortes DM, et al. (2006) Molecular determinants of gating at the potassium-channel selectivity filter. Nat Struct Mol Biol 13: 311–318. doi:10.1038/nsmb1069.
36. Cordero-Morales JF, Jogini V, Lewis A, Vásquez V, Cortes DM, et al. (2007) Molecular driving forces determining potassium channel slow inactivation. Nat Struct Mol Biol 14: 1062–1069. doi:10.1038/nsmb1309.
37. Bhate MP, McDermott AE (2012) Protonation state of E71 in KcsA and its role for channel collapse and inactivation. Proc Natl Acad Sci U S A 2012: 1–6. doi:10.1073/pnas.1211900109.
38. Aqvist J, Luzhkov V (2000) Ion permeation mechanism of the potassium channel. Nature 404: 881–884. doi:10.1038/35009114.
39. Irizarry SN, Kutluay E, Drews G, Hart SJ, Heginbotham L (2002) Opening the KcsA K + Channel: Tryptophan Scanning and Complementation Analysis Lead to Mutants with Altered Gating. Biochemistry 41: 13653–13662. doi:10.1021/bi026393r.
40. Yifrach O, MacKinnon R (2002) Energetics of Pore Opening in a Voltage-Gated K+ Channel. Cell 111: 231–239. doi:10.1016/S0092-8674(02)01013-9.
41. Sadovsky E, Yifrach O (2007) Principles underlying energetic coupling along an allosteric communication trajectory of a voltage-activated K+ channel. Proc Natl Acad Sci U S A 104: 19813–19818. doi:10.1073/pnas.0708120104.
42. Paynter J, Sarkies P, Andres-Enguix I, Tucker SJ (2008) Genetic selection of activatory mutations in KcsA. Channels 2: 413–418. doi:10.4161/chan.2.6.6874.
43. Lee S-Y, Banerjee A, MacKinnon R (2009) Two separate interfaces between the voltage sensor and pore are required for the function of voltage-dependent K(+) channels. PLoS Biol 7: e47. doi:10.1371/journal.pbio.1000047.
44. Jiang Y, Lee A, Chen J, Ruta V, Cadene M, et al. (2003) X-ray structure of a voltage-dependent K+ channel. Nature 423: 33–41. doi:10.1038/nature01580.
45. Long SB, Campbell EB, Mackinnon R (2005) Crystal structure of a mammalian voltage-dependent Shaker family K+ channel. Science 309: 897–903. doi:10.1126/science.1116269.
46. Hackos DH (2002) Scanning the Intracellular S6 Activation Gate in the Shaker K+ Channel. J Gen Physiol 119: 521–532. doi:10.1085/jgp.20028569.
47. Li-Smerin Y, Hackos DH, Swartz KJ (2000) A Localized Interaction Surface for Voltage-Sensing Domains on the Pore Domain of a K+ Channel. Neuron 25: 411–423. doi:10.1016/S0896-6273(00)80904-6.
48. Tytgat J, Hess P (1992) Evidence for cooperative interactions in potassium channel gating. Nature 359: 420–423. doi:10.1038/359420a0.
49. Smith-Maxwell CJ, Ledwell JL, Aldrich RW (1998) Role of the S4 in cooperativity of voltage-dependent potassium channel activation. J Gen Physiol 111: 399–420. doi:10.1085/jgp.111.3.399.
50. Ledwell JL, Aldrich RW (1999) Mutations in the S4 region isolate the final voltage-dependent cooperative step in potassium channel activation. J Gen Physiol 113: 389–414. doi:10.1085/jgp.113.3.389.

51. Pathak M, Kurtz L, Tombola F, Isacoff E (2005) The cooperative voltage sensor motion that gates a potassium channel. J Gen Physiol 125: 57–69. doi:10.1085/jgp.200409197.
52. Blunck R, McGuire H, Hyde HC, Bezanilla F (2008) Fluorescence detection of the movement of single KcsA subunits reveals cooperativity. Proc Natl Acad Sci U S A 105: 20263–20268. doi:10.1073/pnas.0807056106.
53. Zandany N, Ovadia M, Orr I, Yifrach O (2008) Direct analysis of cooperativity in multisubunit allosteric proteins. Proc Natl Acad Sci U S A 105: 11697–11702. doi:10.1073/pnas.0804104105.
54. Zhou Y, Morais-Cabral JH, Kaufman A, MacKinnon R (2001) Chemistry of ion coordination and hydration revealed by a K+ channel-Fab complex at 2.0 Å resolution. Nature 414: 43–48. doi:10.1038/35102009.
55. Guex N, Peitsch MC (1997) SWISS-MODEL and the Swiss-PdbViewer: an environment for comparative protein modeling. Electrophoresis 18: 2714–2723. doi:10.1002/elps.1150181505.
56. Wolf MG, Hoefling M, Aponte-Santamaría C, Grubmüller H, Groenhof G (2010) g_membed: Efficient insertion of a membrane protein into an equilibrated lipid bilayer with minimal perturbation. J Comput Chem 31: 2169–2174. doi:10.1002/jcc.21507.
57. Knape K, Linder T, Wolschann P, Beyer A, Stary-Weinzinger A (2011) In silico Analysis of Conformational Changes Induced by Mutation of Aromatic Binding Residues: Consequences for Drug Binding in the hERG K+ Channel. PloS one 6: e28778. doi:10.1371/journal.pone.0028778.
58. Hess B, Kutzner C, Van der Spoel D, Lindahl E (2008) GROMACS 4: Algorithms for Highly Efficient, Load-Balanced, and Scalable Molecular Simulation. J Chem Theory Comput 4: 435–447. doi:10.1021/ct700301q.
59. Hornak V, Abel R, Okur A, Strockbine B, Roitberg A, et al. (2006) Comparison of multiple Amber force fields and development of improved protein backbone parameters. Proteins 65: 712–725. doi:10.1002/prot.21123.
60. Jorgensen WL, Chandrasekhar J, Madura JD, Impey RW, Klein ML (1983) Comparison of simple potential functions for simulating liquid water. J Chem Phys 79: 926. doi:10.1063/1.445869.
61. Siu SWI, Vácha R, Jungwirth P, Böckmann R a (2008) Biomolecular simulations of membranes: physical properties from different force fields. J Chem Phys 128: 125103. doi:10.1063/1.2897760.
62. Liu Y, Nagle JF (2004) Diffuse scattering provides material parameters and electron density profiles of biomembranes. Phys Rev E: Stat, Nonlinear, Soft Matter Phys 69: 040901. doi:10.1103/PhysRevE.69.040901.
63. Darden T, York D, Pedersen L (1993) Particle mesh Ewald: An N log(N) method for Ewald sums in large systems. J Chem Phys 98: 10089. doi:10.1063/1.464397.
64. Hess B, Bekker H, Berendsen HJC, Fraaije JGEM (1997) AID-JCC4<>3.0.CO;2-H.
65. Amadei A, Linssen AB, Berendsen HJ (1993) Essential dynamics of proteins. Proteins 17: 412–425. doi:10.1002/prot.340170408.
66. Hub JS, De Groot BL, Van der Spoel D (2010) g_wham—A Free Weighted Histogram Analysis Implementation Including Robust Error and Autocorrelation Estimates. J Chem Theory Comput 6: 3713–3720. doi:10.1021/ct100494z.

5.5 Paper #5

In silico Analysis of Conformational Changes Induced by Mutation of Aromatic Binding Residues: Consequences for Drug Binding in the hERG K+ Channel

Kirsten Knape[1], Tobias Linder[2], Peter Wolschann[1], Anton Beyer[1], Anna Stary-Weinzinger[2]*

1 Institute for Theoretical Chemistry, University of Vienna, Vienna, Austria, 2 Department of Pharmacology and Toxicology, University of Vienna, Vienna, Austria

Abstract

Pharmacological inhibition of cardiac hERG K+ channels is associated with increased risk of lethal arrhythmias. Many drugs reduce hERG current by directly binding to the channel, thereby blocking ion conduction. Mutation of two aromatic residues (F656 and Y652) substantially decreases the potency of numerous structurally diverse compounds. Nevertheless, some drugs are only weakly affected by mutation Y652A. In this study we utilize molecular dynamics simulations and docking studies to analyze the different effects of mutation Y652A on a selected number of hERG blockers. MD simulations reveal conformational changes in the binding site induced by mutation Y652A. Loss of π-π-stacking between the two aromatic residues induces a conformational change of the F656 side chain from a cavity facing to cavity lining orientation. Docking studies and MD simulations qualitatively reproduce the diverse experimentally observed modulatory effects of mutation Y652A and provide a new structural interpretation for the sensitivity differences.

Citation: Knape K, Linder T, Wolschann P, Beyer A, Stary-Weinzinger A (2011) *In silico* Analysis of Conformational Changes Induced by Mutation of Aromatic Binding Residues: Consequences for Drug Binding in the hERG K+ Channel. PLoS ONE 6(12): e28778. doi:10.1371/journal.pone.0028778

Editor: Ying Xu, University of Georgia, United States of America

Received May 31, 2011; **Accepted** November 15, 2011; **Published** December 15, 2011

Copyright: © 2011 Knape et al. This is an open-access article distributed under the terms of the Creative Commons Attribution License, which permits unrestricted use, distribution, and reproduction in any medium, provided the original author and source are credited.

Funding: This work was supported by the Wiener Hochschuljubiläumsstiftung (Grant H-2246/2010) and The Austrian Science Fund (FWF; Grants P22395, W1232). The funders had no role in study design, data collection and analysis, decision to publish, or preparation of the manuscript.

Competing Interests: The authors have declared that no competing interests exist.

* E-mail: anna777@gmx.at

Introduction

HERG (human ether-a-go-go related gene) encodes the pore-forming subunit of the voltage-gated potassium channel I_{Kr}, expressed in the heart and in nervous tissue [1]. The channel contributes to modulation of the repolarization phase III of the myocyte action potential [1–3]. Disruption of hERG channel function, due to inherited mutations [4], [5], or side effects of drugs, has been linked to long QT syndrome (LQTS) [6], which may lead to serious arrhythmia and sudden cardiac death [7], [8]. This phenomenon is caused by structurally diverse therapeutic compounds including antiarrhythmics, antihistamines, antipsychotics and antibiotics [9]. Several compounds like terfenadine (Seldane®) and cisapride (Propulsid®) had to be withdrawn from the market for this reason. Consequently, there is an intense interest in understanding the molecular and structural mechanisms of hERG channel gating and block. Individual mutations of pore forming residues to alanine revealed amino acids essential for drug binding. Residues T623, S624 and V625, located at the bottom of the pore helix, and residues G648, Y652 and F656, located in S6 segments are important binding determinants for many drugs from diverse chemical classes [1], [10–21]. Mutations of Y652 and F656 to alanine resulted in 94-fold and 650-fold block decrease for compound MK-499, respectively [10]. Similar strong effects have been found for many structurally unrelated compounds such as cisapride and terfenadine, suggesting a common binding region within the aqueous inner cavity [22].

Homology models [23–27] suggest that high affinity binding determinants Y652 and F656 are arranged in two aromatic rings, facing the inner cavity (Fig. 1). π-π-stacking interactions as well as cation-π-interactions with these residues have been proposed to play a crucial role for block [28]. The importance of the aromatic side chain at position Y652 is further supported by mutational studies, indicating that conservative mutations Y652F and Y652W retain normal sensitivity to high affinity blockers MK-499 and cisapride [10] while non-aromatic substitutions strongly diminish block. In contrast, at position F656 hydrophobicity seems sufficient for high affinity block [16].

The binding mode for blockers such as bepridil, thioridazine or fluvoxamine differs with respect to Y652. These compounds are only partially attenuated by mutation Y652A [22], [28–30]. Nevertheless, with the exception of fluvoxamine [28], drugs are strongly attenuated by mutation F656A, suggesting that they bind in the inner cavity [12], [31]. In 2009, Xing et al. [32] found that capsaicin, a pungent irritant occurring in peppers, enhances hERG block upon mutation of Y652A 4-fold, while F656 was suggested to be relatively unimportant for block.

The mechanism by which these drugs interact with hERG channels is largely unknown. Thus, we investigated whether bepridil, thioridazine, propafenone and capsaicin have different binding modes compared to cisapride, dofetilide, E-4031, MK-499, terfenadine or ibutilide. In this study we utilized MD simulations and docking studies to investigate the different role of Y652 on drug binding.

Results

Flexibility of putative aromatic binding residues in the hERG cavity

A recently validated homology model of the open hERG pore (model 6 of Stary et al. [26]) was used as starting point for our

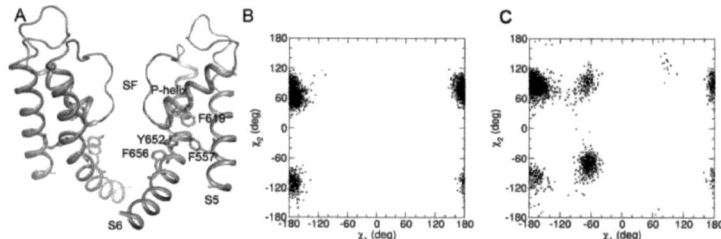

Figure 1. Location and flexibility of putative aromatic binding residues in hERG. (A) π-π-stacking interactions between binding determinants Y652 and F656, located on helix S6 and residues F619 (P-helix) and F557 (S5 helix). Side chains are shown as green sticks (B). χ_1/χ_2 plot of Y652 and F656 (C) obtained from 50 ns MD simulations.
doi:10.1371/journal.pone.0028778.g001

analyses. Y652 and F656 belong to a cluster of four aromatic residues, which includes F557 located on helix S5 and F619 from the P segment (Figure 1A). The conformational flexibility of these aromatic side chains was analyzed using molecular dynamics simulations. Figure 1B–C shows the distribution of dihedral angles χ_1 (rotation around Cα–Cβ atoms) and χ_2 (rotation around Cβ–Cγ atoms) for side chains Y652 and F656 on a 50 ns time scale. Since our sampling protocol involved sampling at 10 ps intervals and each channel contains four homologous domains, each plot contains 20,000 black dots representing the conformations observed in the simulation. Figure 1B illustrates the rigidity of the Y652 side chain on the nanosecond time scale. Variations are observed for the dihedral angle χ_2 only. The F656 side chain is more mobile, it can adopt various χ_1 and χ_2 conformations. The multiple observed conformational states suggest inherited flexibility at position F656 in the open conformation. The side chain of F557, which is not part of the drug binding site, is relatively rigid. The phenyl ring of F619 from the P-helix adopts various χ_1 and χ_2 conformations (see Figure S1A–B).

Conformational changes induced by alanine mutations

The structural effects of mutations Y652A and F656A were examined using MD simulations. First, *in silico* mutants were generated using the mutagenesis tool in PyMOL, followed by energy-minimizations. Repeated simulations on a 50 ns time scale were performed. The stability of the mutant channels, measured as the root mean square deviation (RMSD) as a function of time is shown in Figure 2A. The values for WT and Y652 are in the range of 0.25 nm, the RMSD for the F656A mutant is slightly higher; it reaches 0.3 nm after 50 ns. The increased RMSD is not due to stability differences in S6 helices (Figure 2C) but due to less stable loops connecting S5 and P helix (Figure 2B).

Replacement of the planar aromatic moiety in position Y652 altered the conformation of residue F656, which was stabilized by parallel displaced π-π-stacking interactions in WT. Calculations by Tsuzuki et al. [33], indicate that the energy contribution for this type of aromatic-aromatic interactions is in the range of −1.48 kcal/mol. Due to the loss of these interactions in the Y652A mutant the side chain of F656 rotated away from the pore axis allowing edge to edge shaped π-π-stacking interactions with F557 from the neighboring S5 segment (see Figure 3A–C). The interaction energies of edge to edge π-π-stacking are approximately 1 kcal/mol stronger than parallel displaced π-π-stacking (−2.48 kcal/mol vs. −1.48 kcal/mol [33]). Figure 3D compares the χ_1 angle of side chain F656 in WT and Y652A mutant channels as a function of time. The χ_1 angle is predominantly in

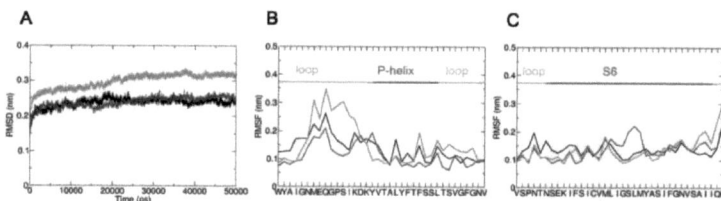

Figure 2. Stability of WT and mutant hERG channels. (A) Backbone RMSD of the Y652A (blue) and the F656A mutant (brown) compared to WT channel (black). (B) Comparison of the root mean square fluctuations (RMSF) for WT and mutant channels. Only the P-helix and connecting loops are shown. (C) RMSF of S6 helix.
doi:10.1371/journal.pone.0028778.g002

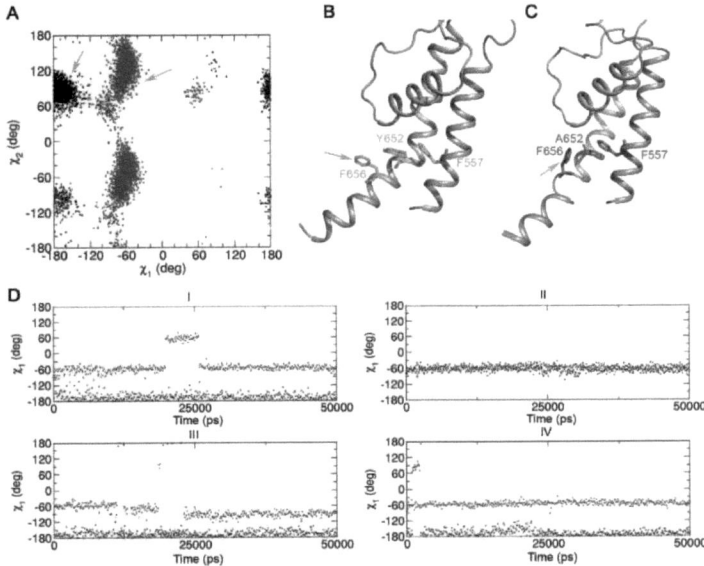

Figure 3. Side chain rearrangements of F656 induced by mutation Y652A. (A) χ_1/χ_2 side-chain angles of F656 for WT (black) and Y652A (blue). The green and blue arrows indicate the approximate conformations of the F656 side chains shown in B and C. (B) Representative side-chain conformations of WT and Y652A mutant (C) channel snapshots taken from MD simulations. (D) F656 χ_1 dihedral angles for WT (black) and Y652A (blue) in all four domains as a function of time.
doi:10.1371/journal.pone.0028778.g003

the range of $-180°$ to $-60°$ in WT (trans orientation). In the mutant channel this value is changed to $-60°$ to $60°$. The more favorable edge to edge stacking energy might explain why the F656 side chain adopted a gauche($-$) conformation in 80% of the simulations. Gauche(+) and trans conformations were rarely observed (results for the rerun are shown in Figure S3).

MD simulations on the F656A mutant did not reveal significant conformational changes of aromatic residues compared to WT (see Figure S1A–C). Therefore, this mutant was not analyzed further.

Docking studies on WT and Y652A mutant channels

We next analyzed the effects of the Y652A mutant induced side chain orientation of residue F656 on drug block. Eleven drugs (Figure 4 and Figure 5) were docked into 20 WT and 20 Y652A snapshots (every 5 ns from two independent runs) derived from 50 ns MD simulations. For each blocker, the ten most frequent occurring docking poses of each drug (n = 100) were analyzed with respect to aromatic ring stacking and/or hydrophobic interactions with binding residues Y652 and F656. Table 1 summarizes these interactions and lists the number of t-shaped (t), edge-to-edge (e) and parallel π-π-stacking (p) interactions. Gold Chemscores (Gold.Chemscore.DG) are listed in Table 2.

Drugs can be divided into three groups according to their binding behavior. For drugs that have been shown to be only partially attenuated by a tyrosine to alanine mutation in position 652 [22], [28–30], no or slight changes in binding behavior compared to WT were observed (Table 1 and Figure 5). The binding mode for thioridazine was identical in WT and Y652A. Three aromatic interactions were predicted in both cases (Figure 5). Docking studies with bepridil suggested that the total number of aromatic interactions remained constant in the mutant channel. However, in the WT channel this drug formed one parallel π-π-stacking interaction with Y652 and one edge-to-edge interaction, while in the Y652A mutant channel, two edge-to-edge interactions with F656 were predicted (Figure 5). The number of aromatic interactions for propafenone and GPV009 did not change in Y652A. The only modification observed was a change of one t-shaped to an edge-to-edge stacking interaction with propafenone.

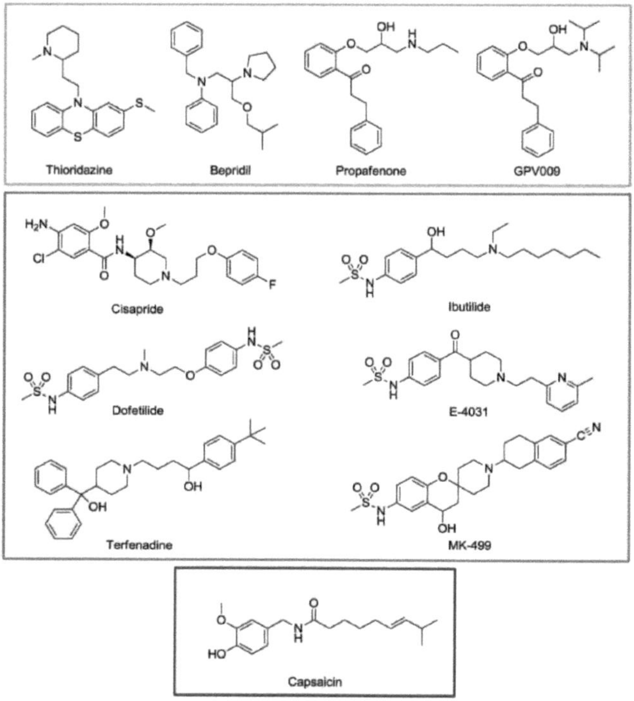

Figure 4. Structures of hERG blockers examined in this study. Drugs are clustered into three groups: group 1 (orange frame) includes blockers which are relatively insensitive to mutation Y652A[22,28–30], group 2 (green frame) shows Y652 sensitive drugs[10,18,22,31] and group 3 (blue frame) shows capsaicin whose affinity is increased by mutation Y652A[32].
doi:10.1371/journal.pone.0028778.g004

In agreement with experimental data, cisapride, dofetilide, E-4031, ibutilide, MK-499 and terfenadine were predicted to strongly interact with aromatic side chains Y652 and F656. While several favorable aromatic interactions to both aromatic side chains were predicted for cisapride and MK-499 in the WT channel, docking studies performed with the Y652A mutant channel indicated complete loss of aromatic interactions. All other drugs in this group had drastically reduced aromatic and hydrophobic interactions with F656 in the Y652A mutant channel. For example, in the WT channel terfenadine was predicted to interact with Y652 side chains from three domains and two F656 residues. In the mutant channel only one edge-to-edge interaction with the F656 side chain remained (see Figure 5 and Figure S2).

Important changes between the Y652A sensitive and Y652A insensitive drug groups were also observed considering the conformation of the drugs. Docking results suggest that thioridazine, bepridil, propafenone and GPV009 fold mostly into U-shaped conformations, while extended conformations parallel to the pore axis were not observed in either WT or Y652A mutant channel. In contrast, most drugs that are highly sensitive to mutation Y652A change their conformation from U-shaped in the WT channel to a stretched conformation longitudinal to the channel axis (Figure 5).

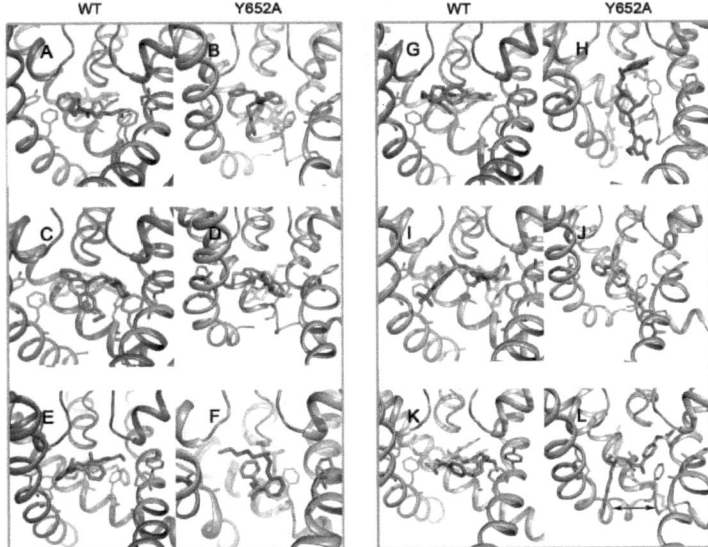

Figure 5. Docking (cyan transparent sticks) and MD poses. at the end of the simulation (blue sticks) of bepridil (AB), thioridazine (CD), propafenone (EF), cisapride (GH), terfenadine (IJ) and ibutilide (KL) in WT and Y652A (from left to right). Y652 and F656 are shown as green lines; A652 is shown as orange lines. The arrow displays the movement of the heptyl chain of ibutilide.
doi:10.1371/journal.pone.0028778.g005

Capsaicin possibly belongs to a third class of drugs, which is affected by the Y652A mutation in a different way. To the best of our knowledge, it is currently the only known drug that shows increased affinity for the Y652A mutant. Gold predicted one aromatic interaction (parallel or edge-to-edge) for WT and Y652A, respectively. Docking suggests that the number of hydrogen bonds with selectivity filter residues T623 and S624 increased in the mutant channel (two H-bonds to T623 and one H-bond to S624). In the WT channel only two hydrogen bonds between capsaicin and S624 were predicted (Figure 6).

MD simulations support different drug binding modes in WT and Y652A channels

To further support our hypothesis, 10 ns MD simulations on all docked poses shown in Figure 5 and Figure S2 were performed. Generally, the docked binding poses are stable on the nanosecond time scale. The Y652A insensitive compounds bepridil, thioridazine, propafenone, and GPV009 retain their compact binding mode in the channel pore in WT and Y652A mutant channels (Figure 5A–F, Figure S2A–B). For the drugs which are sensitive to mutation Y652A, simulations strongly suggest the suggested drug rearrangement from the horizontal binding mode in WT to a stretched conformation along the channel axis in the mutants (Figure 5G–L, Figure S2C–H). This provides a possible explanation for the experimentally observed affinity loss. Only E-4031 does not remain stable in the Y652A mutant (Figure 5J).

Additionally, MD simulations reveal which functional groups of the compounds are flexible. For example, while the basic scaffold of the propafenone molecule (acylphenyloxypropanolamine) remains rather rigid, the side chain adopts various conformations. The movies S1, S2, S3, S4, S5, S6, S7, S8, S9, and S10 show the behavior of all ten drugs in WT and mutant channels during the 10 ns MD simulation runs.

Surprisingly, the conformational flexibility of the Y652 and F656 side chains is not influenced when drugs reside in the cavity (for examples see Figure S6). In contrast, conformational changes of the aromatic side chains sometimes induce changes in drug orientation (for example see behavior of E-4031 in the movie S5).

Discussion

Direct block of hERG channels by structurally diverse drugs is mediated by aromatic side chains Y652 and F656 (see Figure 1 for

Table 1. Aromatic ring stacking and hydrophobic interactions (HIA) between Y652 and F656 side chains and hERG antagonists in WT and Y652A mutant channels.

Compound	WT			Y652A		
	Y652	F656	∑ HIA	A652	F656	∑ HIA
Thioridazine	-	3 (t,p,e)	3	-	3 (t,p,e)	3
Bepridil	1 (p)	1 (e)	2	-	2 (e)	2
Propafenone	2 (t,e)	2 (t,e)	4	-	2 (e)	2
GPV0009	3 (t,2e)	1 (e)	4	-	1 (e)	1
Capsaicin	-	1 (p)	1	-	1 (e)	1
Cisapride	2 (p,e)	2 (t,p)	4	-	-	0
Dofetilide	2 (t,e)	2 (p,e)	4	-	1 (e)	1
E-4031	3 (t,2e)	2 (e)	5	-	2 (p,e)	2
Ibutilide	3 (e)	-	3	-	1 (p)	1
MK-499	1 (e)	1 (e)	2	-	-	0
Terfenadine	3 (t,2e)	2 (t,e)	5	-	1 (e)	1

(t = T-shaped stacking, p = parallel π-π-stacking, e = edge-to-edge interactions).
doi:10.1371/journal.pone.0028778.t001

Table 2. Free energies of binding calculated by Chemscore (ΔG^{bind} kJ/mol) for WT and Y652A.

Drug	Chemscore WT	Chemscore Y652A	Difference WT vs Y652A
Thioridazine	−30.67	−27.83	−2.84
Bepridil	−32.92	−31.23	−1.69
Propafenone	−32.88	−29.75	−3.13
GPV0009	−35.51	−31.43	−4.08
Capsaicin	−31.34	−32.85	1.51
Cisapride	−30.86	−22.96	−7.90
Dofetilide	−30.00	−20.67	−9.33
E-4031	−35.59	−22.82	−12.77
Ibutilide	−33.69	−21.52	−12.17
MK-499	−30.18	−23.19	−6.99
Terfenadine	−35.21	−30.51	−4.70

doi:10.1371/journal.pone.0028778.t002

location of residues) [17]. Mutation of either residue to alanine dramatically reduces drug potency, implying a direct interaction with these residues. In agreement with this hypothesis, various drug docking studies predict binding modes, favoring π-π-stacking interactions with Y652 and F656.

More recently, compounds have been identified, which are insensitive to mutation Y652A, while displaying greatly reduced affinity for the F656A mutant [22], [28], [29]. The lack of sensitivity of these molecules could simply result from binding "less deeply" in the cavity, possibly below the position of Y652. Alternatively, replacement of Y652 might induce allosteric effects on drug binding rather than directly disrupting binding. For the reasons discussed in detail below, we favor the second hypothesis.

Our MD simulations suggest that deletion of the aromatic side chain in position Y652 induces allosteric changes in the drug binding site, with important consequences for drug binding. Specifically, loss of π-π-stacking interactions induce a conformational change of the F656 side chain from a cavity facing orientation (χ_1 values in the range of −180°) to a cavity lining conformation (χ_1 values in the range of −60°) (compare Figure 1A–C and 3A–C). This conformation is stabilized by energetically favorable edge-to-edge stacking interactions with the F557 aryl ring, located on helix S5. Docking studies comparing the binding modes of 11 hERG blockers revealed different behavior for rigid compact molecules versus compounds with more extended geometries. Only the first class of drugs could still favorably interact with the reoriented F656 residues from several subunits in the Y652A mutant (Figure 5, Table 1 and Figure S2). These results correlate well with experimental Y652A sensitivities and are in agreement with a ligand based hypothesis by Stansfeld et al. [34], [35] derived from a study of 20 LQT compounds with varying Y652A sensitivities. Besides, the importance of the orientation of the Y652 and F656 side chains for high affinity block has been elegantly demonstrated by Chen et al. [36] Their study showed that the decreased drug affinity of non-inactivating hERG mutant channels is not caused by inactivation per se but by inactivation gating-associated reorientation of residues located in the S6 domain.

It has been reported by Zachariae et al. [37] that longer molecules bind in a perpendicular orientation to the channel axis and therefore may interact with all four domains of the channel. In our WT drug docking studies we observe the same perpendicular positioning. In the Y652A mutant, the orientation of extended compounds, sensitive to mutation of Y652 is changed to a stretched conformation parallel to the channel axis. These drug reorientations in the Y652A mutant are further supported by a total of 200 ns (10 and for WT and mutants, respectively) MD simulations.

In contrast to an interesting study by Huang et al. [38], who observed an induced fit of a toxin binding to the extracellular side of the selectivity filter in a shaker K$^+$ channel, we did not see conformational adaptions of Y652 or F656 upon drug binding to the hERG inner cavity (Figure S5). This suggests different drug receptor interactions for different binding sites, which might be in part explained by the different nature of interactions (mainly electrostatic versus mainly aromatic/hydrophobic). Interestingly, conformational changes of the aromatic side chains sometimes even induce changes in drug orientation (see movies S1, S2, S3, S4, S5, S6, S7, S8, S9, and S10).

In a recent review by Zhou et al. [39], it was pointed out that aromatic side chains are predestinated to serve as channel gates, preventing ion flow in the closed conformation. Detailed inspection of our recently published closed hERG homology model [40] indeed reveals an optimal arrangement of the F656

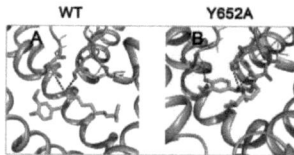

Figure 6. Interactions of capsaicin with the selectivity filter in WT (A) and Y652A mutant (B). Selectivity filter residues involved in capsaicin binding are shown as green sticks; residues of the TSV motif not interacting with capsaicin are shown as grey lines. Hydrogen bonds are depicted as black dots.
doi:10.1371/journal.pone.0028778.g006

side chains to prevent ion flow. In this study, the side chain reorientations in the Y652A mutant most likely do not influence ion conductance in the open state, however we cannot exclude gating-associated reorientations. It was beyond the scope of the current study to analyze mutation induced effects in the closed channel state. Nevertheless, future studies might provide an answer to this important question.

To further support the allosteric side chain rearrangement hypothesis, we introduced a phenylalanine at position Y652, which was shown to restore WT-like binding behavior for the high affinity compound MK-499 [20]. MD simulations show that in the Y652F mutant the aromatic side chains of F557, F619 and F656 behave similar as in the WT channel (see Figure S4A–D). In agreement with experimental data, we find similar docking poses for WT and Y652F channels with compound MK-499.

Allosteric effects on drug block are also used to explain the effects of several inactivation deficient mutants on hERG block. For example, inactivation deficient mutants N588K and S620T exhibit reduced affinity for dofetilide and other high affinity blockers [41]. However, none of these residues is assumed to directly interact with these compounds, since both are located distantly from the binding site.

In conclusion, MD simulations of WT and Y652A mutant channels in combination with drug docking provide a new structural interpretation for the diverse modulatory effects of residue Y652 on different hERG blockers ranging from strong affinity decrease (e.g. cisapride) upon mutation to affinity increase in the case of capsaicin. The results provide a starting point for future investigations focusing on further residues of the aromatic cluster in the hERG binding site. For example, studies on mutants of F557 and F619 will provide a better understanding of the still poorly described mechanisms underlying hERG block.

Materials and Methods

Molecular dynamics simulations

MD simulations were performed with Gromacs v.4.5.4. [42] Two independent simulation setups using either the OPLS-all-atom force field [43] or the amber99sb force field [44] were used to analyze the dynamics of hERG WT and mutant channels. Mutants Y652A, Y652F and F656A were generated using the mutagenesis tool in PyMOL 0.99 [45] In the OPLS setup, hERG WT and mutant channels were embedded in an equilibrated simulation box of 241 palmitoyloleoylphosphatidylcholine (POPC) lipids. The channels were inserted into the membrane as described previously [26]. K^+ ions were placed in the channel at K^+ sites S0, S2, and S4, with waters placed at S1 and S3 of the selectivity filter [46]. Cl^- ions were added randomly within the solvent to neutralize the system. Lipid parameters were taken from Berger et al. [47]. The solvent was described by the TIP4P water model [48]. Electrostatic interactions were calculated explicitly at a distance <1 nm and long-range electrostatic interactions were calculated at every step by particle-mesh Ewald summation [49]. Lennard-Jones interactions were calculated with a cutoff of 1 nm. All bonds were constrained by using the LINCS algorithm [50], allowing for an integration time step of 2 fs. The Nose-Hoover thermostat was used to keep simulation temperature constant by weakly ($\tau = 0.1$ ps) coupling the lipids, protein and solvent (water+counter-ions) separately to a temperature bath of 300 K. The pressure was kept constant by weakly coupling the system to a pressure bath of 1 bar using a semi-isotropic Parrinello-Rahman barostat algorithm with a coupling constant of 1 ps. Prior to simulations, 1000 conjugate gradient energy-minimization steps were performed, followed by 2 ns of restrained MD in which the protein atoms were restrained with a force constant of 1000 kJ/mol^{-1} nm^{-2} to their initial position. Ions, lipids and solvent were allowed to move freely during equilibration. The systems were then subjected to 50 ns (15 ns Y652F) of unrestrained MD, during which coordinates were saved every 10 ps for analysis. Residues at the N- and C-termini were considered as uncharged, as neither lie at the actual termini of the complete channel. In the amber99sb setup hERG WT and mutant channels were embedded in an equilibrated membrane consisting of 280 dioleoylphosphatidylcholine (DOPC) lipids. Lipid parameters were taken from Siu, et al. [51] and the TIP3P water model [48] was utilized. All further parameters and steps were carried out as described above. Drug topologies were generated using antechamber, which is part of the Amber 11 program package [52]. Charges were taken from Gaussian runs described in the docking section below. After energy-minimization (1000 conjugate gradient energy-minimization steps), unrestrained 10 ns MD simulations for each compound were carried out for WT and Y652A (200 ns in total) at 310 K.

Drug docking

Coordinates of the drugs were generated with Gaussview 5 [53] and the geometry optimized with HF/3-21G implemented in Gaussian09 [53]. For thioridazine, propafenone, GPV009, terfenadine, MK-499 and ibutilide (R)- and (S)-conformations were docked. As no differences could be observed between both enantiomers, only the (R)-conformation was used for further analysis. Docking was performed with the program Gold 4.0.1 [54] using the Gold and Chemscore scoring functions. The coordinates of the geometric center calculated among the Y652 and F656 residues were taken as binding site origin. The binding site radius was set equal to 10 Å. 100,000 operations of the GOLD genetic algorithm were used to dock the selected compounds into the WT and mutant channels. Snapshots after 8, 10, 15, 20, 25, 30, 35, 40, 45 and 50 ns were taken from our 50 ns or 15 ns (Y652F) MD trajectories. The best ranked 100 poses of each docking run were used for visual analysis of binding. From the ten most occurring positions the numbers of aromatic interactions were averaged.

Supporting Information

Figure S1 χ_1/χ_2 **plots for F557 (A), F619 (B) and Y652 (C) in hERG WT channel (black) and F656A (brown).**
(TIF)

Figure S2 **GPV009 (AB), MK-499 (CD), E-4031 (EF) and dofetilide (GH) in WT and Y652A (from left to right).** Cyan transparent sticks show the docking pose and blue sticks the MD pose and the end of the simulation. The black arrow indicates the moving direction of E-4031 (the dynamical movement of the drug can be observed in the attached movie).
(TIF)

Figure S3 **MD simulation rerun (50 ns) for Y652A mutant. RMSD plot (A), χ_1/χ_2 plot for F656 (B) and the F656 χ_1 dihedral angles in all four domains as a function of time (C) show no significant deviation from the original run.**
(TIF)

Figure S4 χ_1/χ_2 **plots for F557 (A), F619 (B), Y/F652 (C) and F656 (D) in hERG WT channel (black) and Y652F (green).** The 50 ns MD simulation shows that the flexibility of the aromatic side chains in the mutant is comparable to the WT channel.
(TIF)

Figure S5 Y652 (A) and F656 (B) χ₁ dihedral angles as a function of time for WT channel without ligand (black) and with bound bepridil (blue) and dofetilide (red). C shows the χ₁ dihedral angles of F656 in the Y652A mutant as a function of time.
(TIFF)

Movie S1 Behavior of docked drug cisapride in a 10 ns MD simulation. In the first part of the movie, the drug behavior in the WT is shown, followed by the Y652A mutant simulation. Y652 and F656 are shown as green lines; A652 is shown as orange lines.
(WMV)

Movie S2 Behavior of docked drug terfenadine in a 10 ns MD simulation. In the first part of the movie, the drug behavior in the WT is shown, followed by the Y652A mutant simulation. Y652 and F656 are shown as green lines; A652 is shown as orange lines.
(WMV)

Movie S3 Behavior of docked drug ibutilide in a 10 ns MD simulation. In the first part of the movie, the drug behavior in the WT is shown, followed by the Y652A mutant simulation. Y652 and F656 are shown as green lines; A652 is shown as orange lines.
(WMV)

Movie S4 Behavior of docked drug MK-499 in a 10 ns MD simulation. In the first part of the movie, the drug behavior in the WT is shown, followed by the Y652A mutant simulation. Y652 and F656 are shown as green lines; A652 is shown as orange lines.
(WMV)

Movie S5 Behavior of docked drug E-4031 in a 10 ns MD simulation. In the first part of the movie, the drug behavior in the WT is shown, followed by the Y652A mutant simulation. Y652 and F656 are shown as green lines; A652 is shown as orange lines.
(WMV)

Movie S6 Behavior of docked drug dofetilide in a 10 ns MD simulation. In the first part of the movie, the drug behavior in the WT is shown, followed by the Y652A mutant simulation. Y652 and F656 are shown as green lines; A652 is shown as orange lines.
(WMV)

Movie S7 Behavior of docked drug bepridil in a 10 ns MD simulation. In the first part of the movie, the drug behavior in the WT is shown, followed by the Y652A mutant simulation. Y652 and F656 are shown as green lines; A652 is shown as orange lines.
(WMV)

Movie S8 Behavior of docked drug thioridazine in a 10 ns MD simulation. In the first part of the movie, the drug behavior in the WT is shown, followed by the Y652A mutant simulation. Y652 and F656 are shown as green lines; A652 is shown as orange lines.
(WMV)

Movie S9 Behavior of docked drug propafenone in a 10 ns MD simulation. In the first part of the movie, the drug behavior in the WT is shown, followed by the Y652A mutant simulation. Y652 and F656 are shown as green lines; A652 is shown as orange lines.
(WMV)

Movie S10 Behavior of docked drug GPV0009 in a 10 ns MD simulation. In the first part of the movie, the drug behavior in the WT is shown, followed by the Y652A mutant simulation. Y652 and F656 are shown as green lines; A652 is shown as orange lines.
(WMV)

Acknowledgments

The computational results presented have been achieved using the Vienna Scientific Cluster (VSC).

Author Contributions

Conceived and designed the experiments: KK TL AS-W. Performed the experiments: KK TL AS-W. Analyzed the data: KK TL AS-W. Contributed reagents/materials/analysis tools: PW AB AS-W. Wrote the paper: KK TL PW AB AS-W.

References

1. Sanguinetti MC, Jiang C, Curran ME, Keating MT (1995) A mechanistic link between an inherited and an acquired cardiac arrhythmia: HERG encodes the IKr potassium channel. Cell 81: 299–307.
2. Jurkiewicz NK, Sanguinetti MC (1993) Rate-dependent prolongation of cardiac action potentials by a methanesulfonanilide class III antiarrhythmic agent. Specific block of rapidly activating delayed rectifier K+ current by dofetilide. Circ Res 72: 75–83.
3. Tseng GN (2001) I(Kr): the hERG channel. J Mol Cell Cardiol 33: 835–849.
4. Haverkamp W, Breithardt G, Camm AJ, Janse MJ, Rosen MR, et al. (2000) The potential for QT prolongation and proarrhythmia by non-antiarrhythmic drugs: clinical and regulatory implications. Report on a policy conference of the European Society of Cardiology. Eur Heart J 21: 1216–1231.
5. Vandenberg J, Walker B, Campbell T (2001) HERG K+ channels: friend and foe. Trends Pharmacol Sci 22: 240–246.
6. Curran ME, Splawski I, Timothy KW, Vincent GM, Green ED, et al. (1995) A molecular basis for cardiac arrhythmia: HERG mutations cause long QT syndrome. Cell 80: 795–803.
7. Chiang CE, Roden DM (2000) The long QT syndromes: genetic basis and clinical implications. J Am Coll Cardiol 36: 1–12.
8. Keating MT, Sanguinetti MC (2001) Molecular and cellular mechanisms of cardiac arrhythmias. Cell 104: 569–580.
9. Fermini B, Fossa AA (2003) The impact of drug-induced QT interval prolongation on drug discovery and development. Nat Rev Drug Discov 2: 439–447.
10. Mitcheson JS, Chen J, Lin M, Culberson C, Sanguinetti MC (2000) A structural basis for drug-induced long QT syndrome. Proc Natl Acad Sci U S A 97: 12329–12333.
11. Heginbotham L, Lu Z, Abramson T, MacKinnon R (1994) Mutations in the K+ channel signature sequence. Biophys J 66: 1061–1067.
12. Sanchez-Chapula JA, Navarro-Polanco RA, Culberson C, Chen J, Sanguinetti M (2002) Molecular determinants of voltage-dependent human ether-a-go-go related gene (HERG) K+ channel block. J Biol Chem 277: 23587–23595.
13. Mitcheson JS, Chen J, Sanguinetti MC (2000) Trapping of a methanesulfonanilide by closure of the HERG potassium channel activation gate. J Gen Physiol 115: 229–240.
14. Lees-Miller JP, Duan Y, Teng GQ, Duff HJ (2000) Molecular determinant of high-affinity dofetilide binding to HERG1 expressed in Xenopus oocytes: involvement of S6 sites. Mol Pharmacol 57: 367–374.
15. Sanchez-Chapula JA, Navarro-Polanco RA, Sanguinetti MC (2004) Block of wild-type and inactivation-deficient human ether-a-gogo-related gene K+ channels by halofantrine. Naunyn Schmiedebergs Arch Pharmacol 370: 484–491.
16. Fernandez D, Ghanta A, Kauffman GW, Sanguinetti MC (2004) Physicochemical features of the hERG channel drug binding site. J Biol Chem 279: 10120–10127.
17. Ridley JM, Dooley PC, Milnes JT, Witchel HJ, Hancox JC (2004) Lidoflazine is a high affinity blocker of the HERG K+ channel. J Mol Cell Cardiol 36: 701–705.
18. Perry M, de Groot MJ, Helliwell R, Leishman D, Tristani-Firouzi M, et al. (2004) Structural determinants of HERG channel block by clofilium and ibutilide. Mol Pharmacol 66: 240–249.
19. Guo J, Gang H, Zhang S (2006) Molecular determinants of cocaine block of human ether-a-go-go-related gene potassium channels. J Pharmacol Exp Ther 317: 865–874.

20. Sanchez-Chapula JA, Ferrer T, Navarro-Polanco RA, Sanguinetti MC (2003) Voltage-dependent profile of human ether-a-go-go-related gene channel block is influenced by a single residue in the S6 transmembrane domain. Mol Pharmacol 63: 1051–1058.
21. Sanguinetti MC, Mitcheson JS (2005) Predicting drug-hERG channel interactions that cause acquired long QT syndrome. Trends Pharmacol Sci 26: 119–124.
22. Kamiya K, Niwa R, Morishima M, Honjo H, Sanguinetti MC (2008) Molecular determinants of hERG channel block by terfenadine and cisapride. J Pharmacol Sci 108: 301–307.
23. Lees-Miller JP, Subbotina JO, Guo J, Yarov-Yarovoy V, Noskov SY, et al. (2009) Interactions of H562 in the S5 helix with T618 and S621 in the pore helix are important determinants of hERG1 potassium channel structure and function. Biophys J 96: 3600–3610.
24. Subbotina J, Yarov-Yarovoy V, Lees-Miller J, Durdagi S, Guo J, et al. (2010) Structural refinement of the hERG1 pore and voltage-sensing domains with ROSETTA-membrane and molecular dynamics simulations. Proteins 78: 2922–2934.
25. Durdagi S, Subbotina J, Lees-Miller J, Guo J, Duff HJ, et al. (2010) Insights into the molecular mechanism of hERG1 channel activation and blockade by drugs. Curr Med Chem 17: 3514–3532.
26. Stary A, Wacker SJ, Boukharta L, Zachariae U, Karimi-Nejad Y, et al. (2010) Toward a consensus model of the HERG potassium channel. ChemMedChem 5: 455–467.
27. Durdagi S, Duff HJ, Noskov SY (2011) Combined receptor and ligand-based approach to the universal pharmacophore model development for studies of drug blockade to the hERG1 pore domain. J Chem Inf Model 51: 463–474.
28. Milnes JT, Crociani O, Arcangeli A, Hancox JC, Witchel HJ (2003) Blockade of HERG potassium currents by fluvoxamine: incomplete attenuation by S6 mutations at F656 or Y652. Br J Pharmacol 139: 887–898.
29. Windisch A, Timin E, Schwarz T, Stork-Riedler D, Erker T, et al. (2011) Trapping and dissociation of propafenone derivatives in HERG channels. Br J Pharmacol 62: 1542–1552.
30. Mitcheson JS (2003) Drug binding to HERG channels: evidence for a 'non-aromatic' binding site for fluvoxamine. Br J Pharmacol 139: 883–884.
31. Kamiya K, Niwa R, Mitcheson JS, Sanguinetti MC (2006) Molecular determinants of HERG channel block. Mol Pharmacol 69: 1709–1716.
32. Xing J, Ma J, Zhang P, Fan X (2010) Block effect of capsaicin on hERG potassium currents is enhanced by S6 mutation at Y652. Eur J Pharmacol 630: 1–9.
33. Tsuzuki S, Honda K, Uchimaru T, Mikami M, Tanabe K (2002) Origin of attraction and directionality of the pi/pi interaction: model chemistry calculations of benzene dimer interaction. J Am Chem Soc 124: 104–112.
34. Mitcheson JS (2008) hERG potassium channels and the structural basis of drug-induced arrhythmias. Chem Res Toxicol 21: 1005–1010.
35. Stansfeld PJ, Gedeck P, Gosling M, Cox B, Mitcheson JS, et al. (2007) Drug block of the hERG potassium channel: insight from modeling. Proteins 68: 568–580.
36. Chen J, Seebohm G, Sanguinetti MC (2002) Position of aromatic residues in the S6 domain, not inactivation, dictates cisapride sensitivity of HERG and eag potassium channels. Proc Natl Acad Sci U S A 99: 12461–12466.
37. Zachariae U, Giordanetto F, Leach AG (2009) Side chain flexibilities in the human ether-a-go-go related gene potassium channel (hERG) together with matched-pair binding studies suggest a new binding mode for channel blockers. J Med Chem 52: 4266–4276.
38. Huang N, Dong F, Zhou HX (2005) Electrostatic recognition and induced fit in the kappa-PVIIA toxin binding to Shaker potassium channel. J Am Chem Soc 127: 6836–6849.
39. Zhou HX, McCammon JA (2010) The gates of ion channels and enzymes. Trends Biochem Sci 35: 179–185.
40. Garg V, Stary-Weinzinger A, Sachse F, Sanguinetti MC (2011) Molecular determinants for activation of human ether-a-go-go-related gene 1 potassium channels by 3-Nitro-N-(4-phenoxyphenyl) benzamide. Mol Pharmacol 80: 630–637.
41. Perrin MJ, Kuchel PW, Campbell TJ, Vandenberg JI (2008) Drug binding to the inactivated state is necessary but not sufficient for high-affinity binding to human ether-a-go-go-related gene channels. Mol Pharmacol 74: 1443–1452.
42. Hess B, Kutzner C, van der Spoel D, Lindahl E (2008) GROMACS 4: algorithms for highly efficient, load-balanced, and scalable molecular simulation. J Chem Theory Comput 4: 435–447.
43. Jorgensen WL, Maxwell DS, Tirado-Rives J (1996) Development and testing of the OPLS all-atom force field on conformational energetics and properties of organic liquids. J Am Chem Soc 118: 11225–11236.
44. Hornak V, Abel R, Okur A, Strockbine B, Roitberg A, et al. (2006) Comparison of multiple Amber force fields and development of improved protein backbone parameters. Proteins 15: 712–725.
45. The PyMOL Molecular Graphics System, Version 1.3, Schrödinger, LLC.
46. Aqvist J, Luzhkov V (2000) Ion permeation mechanism of the K+ channel. Nature 404: 881–884.
47. Berger O, Edholm O, Jähnig F (1997) Molecular dynamics simulations of a fluid bilayer of dipalmitoylphosphatidylcholine at full hydration, constant pressure, and constant temperature. Biophys J 72: 2002–2013.
48. Jorgensen WL, Chandrasekhar J, Madura JD, Impey RW, Klein ML (1983) Comparison of simple potential functions for simulating liquid water. J Chem Phys 79: 926–935.
49. Darden T, York D, Pedersen L (1993) Particle Mesh Ewald: an NLog(N) method for Ewald sums in large systems. J Chem Phys 98: 10089–10092.
50. Hess B, Bekker H, Berendsen HJC, Fraaije JGEM (1997) LINCS: a linear constraint solver for molecular simulations. J Comput Chem 18: 1463–1472.
51. Siu SW, Vächa R, Jungwirth P, Böckmann RA (2008) Biomolecular simulations of membranes: physical properties from different force fields. J Chem Phys 128: 125103.
52. Case DA, Darden TA, Cheatham TE, Simmerling CL, Wang J, et al. (2010) AMBER 11, University of California, San Francisco.
53. Frisch MJ, Trucks GW, Schlegel HB, Scuseria GE, Robb MA, et al. (2009) Gaussian 09, Revision A.1 Gaussian, Inc., Wallingford CT.
54. GOLD, version 4.0; Cambridge Crystallographic Data Centre: Cambridge, U.K.

5.6 Paper #6

Structural Insights into Trapping and Dissociation of Small Molecules in K⁺ Channels

Tobias Linder, Priyanka Saxena, Eugen Timin, Steffen Hering, and Anna Stary-Weinzinger*

Department for Pharmacology and Toxicology, University of Vienna, Althanstraße 14, 1090 Vienna, Austria

Supporting Information

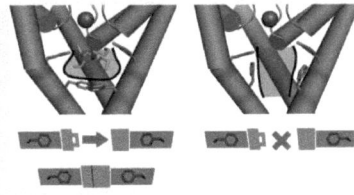

ABSTRACT: K⁺ channels play a critical role in numerous physiological and pathophysiological processes rendering them an attractive target for therapeutic intervention. However, the hERG K⁺ channel poses a special challenge in drug discovery, since block of this channel by a plethora of diverse chemical entities can lead to long QT syndrome and sudden death. Of particular interest is the so-called trapping phenomenon, characterized by capture of a drug behind closed channel gates, which harbors an increased pro-arrhythmic risk. In this study we investigated the influence of trapped blockers on the gating dynamics and probed the state dependence of dissociation in K⁺ channels by making use of the quaternary tetrabutylammonium. By applying essential dynamics simulations and two-electrode voltage clamp we obtained detailed insights into the dynamics of trapping in KcsA and hERG. Our simulations suggest that the trapped TBA influences the F656 flexibility during gate closure. Based on these findings, we provide a structural hypothesis for drug trapping. Further our simulations reveal the extent of gate opening necessary for drug dissociation.

■ INTRODUCTION

Voltage gated K⁺ channels play a critical role in numerous physiological and pathophysiological processes such as nerve and muscle excitation, sensory transduction, and cell proliferation. With a wide range of human diseases linked to voltage gated K⁺ channels, so-called "channelopathies", they represent an attractive target for therapeutic intervention.[1] The human ether-à-go-go related gene (hERG) K⁺ channel poses a special challenge in drug discovery, since it can be blocked by a plethora of structurally diverse drugs including antiarrhythmics, antihistamines, antipsychotics, and antibiotics.[2] This often unwanted inhibition can lead to acquired long QT syndrome (LQTS) and sudden cardiac death.[3,4] Consequently, several pharmaceuticals such as cisapride or terfenadine were withdrawn from the market. Further, reduced hERG channel function caused by inherited mutation leads to congenital LQTS. Recent research indicates that hERG channels are frequently overexpressed in certain human cancers and that long-term treatment with blockers could have therapeutic potential in cancer treatment.[5,6] Thus, pharmacological intervention could potentially be useful for cancer treatment and clinical management of LQTS.

Great efforts have been directed toward a better understanding of the molecular and structural mechanisms of hERG channel gating and block. Substantial progress has been made by identifying the amino acids essential for drug block. They include T623, S624, and V625, from the pore helix, and residues G648, Y652, and F656, located on the TM2 segments.[7-20] However, a key unresolved question in hERG channel block is how drug dissociation is influenced by channel closure. There is evidence that hERG channel blockers can be trapped in the inner cavity of the closed channel.[14,15,18,20-24] The importance of this phenomenon is emphasized by a recent study highlighting a connection between pro-arrhythmic risk and trapping.[25]

The trapping phenomenon in K⁺ channels was first described for quaternary ammonium (QA) blockers by Armstrong in 1971.[26] Since then, QA compounds have been widely used as structural probes to identify the binding side of ion channel blockers,[27-30] to investigate gating processes,[31,32] and to shed light on the structure of ion channels.[33-36]

Herein, we set out to investigate the influence of trapped blockers on the gating dynamics and probe the state dependence of dissociation in K⁺ channels by making use of the QA blocker tetrabutylammonium (TBA). The crystal structure complex of the prototypical K⁺ channel KcsA with a TBA bound in the closed channel pore[37,38] provides an excellent starting point to analyze trapping dynamics and drug dissociation pathways in K⁺ channels. We utilized essential dynamics (ED) simulations and two-electrode voltage clamp to obtain detailed insights into the dynamics of trapping in KcsA and hERG. Further, free energy calculations were performed to examine state dependent dissociation of a trapped channel blocker.

Received: June 16, 2014
Published: October 9, 2014

■ METHODS

MD Simulations. The crystal structure (pdb identifier: 2HVK), comprising of the closed KcsA channel from residue 22 to 124 and the cocrystallized trapped TBA,[38] was used as a starting point for MD simulations. The open (pdb identifier: 3f7v) and the intermediate (pdb identifier: 3lb6) channel states were obtained from Cuello et al.[39] Due to crystallization of shorter fragments in the latter two structures, the missing amino acids at the N- and C-termini below the bundle crossing gate were added using Discovery Studio 3.5 (Accelrys Software Inc., San Diego, CA, USA) to obtain channel states of the same length. In addition, the introduced cysteine at position 90 in the crystal structures was mutated back to the WT leucine. TBA was added to the open and intermediate channels by placing it in the cavity according to the closed channel binding site. Further, TBA was docked to the open hERG homology model[40] using FlexX which is part of the LeadIT software package version 2.0.1 (BioSolveIT, Sankt Augustin, Germany). The sphere center of the binding site was placed in the middle of the cavity, and the radius was set to 10 Å. General amber force field parameters[41] for TBA were generated by making use of Gaussian09[42] and antechamber[43] which is part of the amber package.[44] Protein−ligand complexes were embedded in an equilibrated palmitoyloleoylphosphatidylcholine (POPC) membrane consisting of 256 lipids using the g_membed tool.[45] K^+ ions were placed in the selectivity filter (SF) at sites S0, S2, and S4, and water was added at S1 and S3.[46] The system was neutralized by randomly adding Cl^- within the solvent. All simulations were carried out using the MD simulation software Gromacs v.4.5.4.[47] The amber99sb force field[48] and the TIP3P water model[49] were employed. Lipid parameters were taken from Berger et al.[50] The cutoff for Lennard-Jones interactions was set to 1.0 nm, and parameters were corrected for monovalent ions.[51] Electrostatic interactions were calculated with a cutoff of 1.0 nm, and long-range electrostatic interactions were treated by the particle-mesh Ewald method at every step.[52] The LINCS algorithm[53] was used to constrain bonds, allowing for an integration step of 2 fs. The Nose-Hoover thermostat[34,55] was used keeping the simulation temperature constant at 310 K. Coupling groups were defined as the protein−ligand complex, lipids, and solvent with a time constant of 0.2 ps. The Parrinello−Rahman barostat algorithm[56] with a coupling constant of 1 ps was used for a constant pressure of 1 bar. Prior to simulation, 1000 conjugate gradient energy-minimization steps and a 5 ns equilibration run by restraining the protein−ligand complex by a force constant of 1000 kJ/mol/nm² were performed. Subsequent free MD simulations were carried out for 20 ns in KcsA and for 100 ns in hERG.

Essential Dynamics Simulations. The ED technique was employed as described previously.[57] Briefly, an eigenvector representing the transition between open and closed channel state was obtained from principal component analysis. For this analysis, the backbone of the helices between the two states was compared. Fixed increment linear expansion was set to $-1.69e^{-6}$ nm per simulation step (2 fs). Twenty closing ED simulations, lasting for 20 ns each, were performed in the absence and presence of TBA for KcsA and hERG channels, respectively.

Force Probe MD Simulations and Umbrella Sampling. To probe different dissociation pathways of TBA, a harmonic potential with a force constant of 1000 kJ/mol/nm² was applied on TBA. The compound was pulled for 20 ns with a rate of 0.00025 nm/ps along the z-axis to investigate dissociation through the activation gate. The first turn of the extracellular side of the P-helix was restrained with a force constant of 1000 kJ/mol/nm² during pulling to prohibit movement of the protein. For dissociation simulations of the open KcsA and hERG channels, ions were restrained with a force constant of 10000 kJ/mol/nm² to prevent ion migration through the gate and minimize the ion influence on drug dissociation. In case of dissociation experiments on KcsA, 95, 99, and 56 snapshots from closed, intermediate, and open state simulations were extracted, respectively, and subject to 20 ns umbrella sampling with force constants of either 1000 or 10000 kJ/mol/nm². For hERG dissociation, 57 snapshots from open simulations were obtained and used for umbrella sampling accordingly (histograms of umbrella sampling simulations are shown in the Supporting Information). The first 10 ns of each window were discarded for equilibration. The potential of mean force and the statistical errors were estimated by the g_wham tool and the integrated bootstrap analysis method using 100 bootstraps.[58]

Experimental Procedure. cDNAs of hERG (accession number NP000229) were kindly provided by Prof. Sanguinetti (University of Utah, UT, USA). Synthesis of capped runoff complementary ribonucleic acid (cRNA) transcripts from linearized cDNA (cDNA) templates and injection of cRNA were performed as described in detail by Sanguinetti et al.[59] Oocytes from the South African clawed frog, *Xenopus laevis* (NASCO, Fort Atkinson, WI, USA), were prepared as follows: After 15 min exposure of female *Xenopus laevis* to the anesthetic (0.2% solution of MS-222; the methanesulfonate salt of 3-aminobenzoic acid ethyl ester; Sigma), parts of the ovary tissue were surgically removed. Defolliculation was achieved by enzymatical treatment with 2 mg/mL collagenase type 1A (Sigma) and mechanical removal of follicular layer using forceps. Stage V–VI oocytes were selected and injected with the WT and mutant hERG-encoding cRNA. Injected oocytes were stored at 18 °C in ND96 bath solution (96 mM sodium chloride, 2 mM potassium chloride, 1 mM magnesium chloride, 5 mM HEPES, 1.8 mM $CaCl_2$; pH 7.5, titrated with NaOH) containing 1% penicillin-streptomycin solution. All chemicals used were purchased from Sigma-Aldrich Chemie GmbH, Taufkirchen, Germany.

Currents through hERG channels were studied 1 to 4 days after microinjection of the cRNA using the two-microelectrode voltage clamp technique. ND96 was used as extracellular recording solution. Voltage-recording and current-injecting microelectrodes were filled with 3 M KCl and had resistances between 0.3 and 2 MΩ. Endogenous currents (estimated in oocytes injected with DEPC water) did not exceed 0.15 μA. Currents >5 μA were discarded to minimize voltage clamp errors. Ionic currents were recorded with a Turbo Tec 03X Amplifier (npi electronic, GmbH, Tamm, Germany) and digitized with a Digidata 1322A (Axon Instruments Inc., Union City, CA, USA). The pClamp software package version 9.2 (Axon Instruments Inc.) was used for data acquisition. Microcal Origin 7.0 was employed for analysis and curve fitting.

A precondition for all measurements was the achievement of stable peak current amplitudes over periods of 10 min after an initial run-up period. A frequency of 0.3 Hz was used for all voltage clamp experiments. Drugs were applied by means of a perfusion system enabling solution exchange within 100 ms.[60] The oocytes were kept for 5 min at a holding potential of −100 mV to equilibrate drug diffusion. The tail current was measured

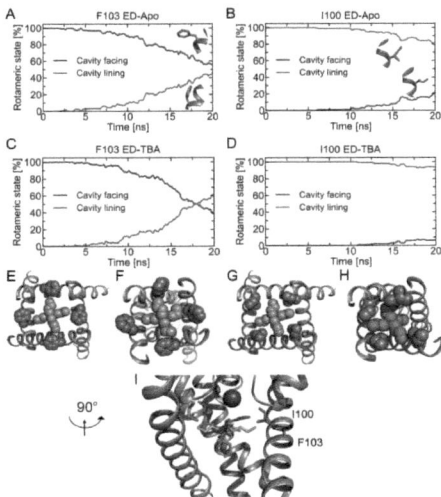

Figure 1. Rotameric states of F103 and I100 during KcsA gate closure. Conformational changes of F103 (A and C) and I100 (B and D) in the absence (A and B) and presence (C and D) of TBA during closure. The cavity facing (χ_1 angle < −123°) and cavity lining (χ_1 angle > −123°) states are shown as blue and green lines, respectively. Panels E and F show the rotameric states of F103 as spheres at the beginning and end of gate closure in top view, respectively. In panels G and H, I100 is shown as spheres accordingly. Panel I represents the xy-plane (pink sticks) and tilted (magenta sticks) orientation of TBA in the side view. For clarity, only three SUs are shown in gray. The K$^+$ ion in the SF is represented as a violet sphere. The color code of the amino acids corresponds to the rotameric states.

at −50 mV, after a step to +20 mV. Use-dependent hERG channel block was estimated as peak tail current inhibition. Data are presented as means ± s.e. from at least four oocytes from ≥2 batches; statistical significance of differences was defined as $P < 0.0001$ in Student's unpaired t-test. The studied compound TBA was obtained from Sigma and was dissolved in ND96 extracellular recording solution to prepare a 1 M stock on the day of experiments. Drug stock solution was further diluted to the required concentration.

■ **RESULTS**

TBA Trapping Simulations in KcsA. Using MD simulations, we have previously identified key residues in KcsA essential for gating.[57] In particular, we found that the TBA binding determinant F103 changes its rotameric state during channel gating. Thus, to investigate the influence of TBA on conformational changes upon channel closure, 20 ED simulations with and without TBA present in the cavity were performed which enable simulating channel closure on the ns time scale.[57] Each closing simulation was conducted for 20 ns. ED is a free MD simulation method, with all coordinates equilibrating, except for one coordinate, which is derived from a linear interpolation between the open and closed helix backbone structures and is biased to drive gating. All other degrees of freedom are explicitly allowed to relax continuously, enabling investigation of side chain conformational changes during drug trapping.

Specifically, the χ_1 angle of the binding site forming F103 and I100 was monitored and separated into the two possible rotameric states, namely cavity lining (χ_1 angle > −123°) and cavity facing (χ_1 angle < −123°) state. Subsequently, the percentage of the two states was calculated for each time step and plotted over time (Figure 1). In the absence of TBA (Figure 1A), nearly equal distribution between the two states was observed at the end of gate closure for F103, which is in good agreement with our previous work.[57] In the presence of TBA, only a slightly faster rotameric switch from cavity facing to the cavity lining was observed over time leading to a preference of the cavity lining conformation at the end of gate closure (Figure 1C). The second important binding determinant, I100, displayed high stability during gating. Only rare changes from the cavity lining to the cavity facing conformation were detected with negligible impact by the bound TBA (Figure 1B and D).

Crystallographic data in combination with MD simulations revealed that TBA can adopt two different orientations in the KcsA binding site linked to the ion configuration in the SF.[37]

Throughout gate closure, TBA remained in the high affinity binding site formed by F103 and I100 (Figure 1E–H). TBA is either in an xy-plane orientation which is parallel to the membrane plane and is centered on the channel symmetry axis. In this orientation, the butyl side chains project into the grooves formed by I100 and F103. Alternatively, TBA adopts a tilted, vertical orientation which is indicated by an off-axis center of TBA (Figure 1I). In the gating simulations, sampling of the xy-plane and the tilted orientation was observed independent of the ion configuration in the SF. No specific pattern between F103 switch and TBA conformation was found indicating that F103 can either face or line the cavity and still allow both TBA orientations. In the rare observations where I100 faced the cavity, spatial displacement led to the tilted orientation of TBA (Figure 1H).

TBA primarily exists in two conformations. The energetically more favorable D_{2d} conformation exhibits a cross-shaped structure with all four butyl chains in a planar arrangement, while the S_4 conformation exhibits the shape of an inverted tetrahedron. Quantum-mechanical calculations of the QA analogue tetraethylammonium (TEA) have shown that the D_{2d} state is separated from the S_4 conformation by energy barriers of around 10 kcal/mol and a total energy difference of around 1 kcal/mol.[61] The energy difference between the two states is similar for TBA.[62] Throughout the closing process, TBA stayed in the D_{2d} conformation (see supplemental Figure S1), which is in good agreement with previous TBA simulations in the closed channel.[37] Only rare transitions to the S_4 conformation were observed. This suggests that TBA does not have to change its conformation in order to allow rotation of the F103 side chain. In addition, this observation indicates that the TBA conformation is independent of the rotameric state of F103. Taken together, our data suggests that TBA does not interfere with side chain rearrangements necessary for gating in KcsA.

Experimental Characterization of hERG Channel Block by TBA. Despite the crucial role of drug trapping in hERG channels[25] the structural interplay between drug and channel during closure are not well understood. Again, we resorted to the well-studied model drug TBA to investigate trapping in hERG. It was previously shown by Choi et al.,[63] that externally applied TBA presumably blocks hERG from the intracellular side by permeating through the cell membrane in CHO cells. However, the characteristics of TBA block were not investigated in detail. Therefore, we set out to test whether TBA is trapped in hERG channels expressed in *Xenopus* oocytes by using the two-electrode voltage-clamp technique.

hERG channels were activated and subsequently inactivated by 300 ms depolarization to +20 mV (Figure 2A). Upon repolarization to −50 mV, channels undergo rapid recovery from inactivation inducing a large tail current. In order to analyze state-dependent block, currents were measured in the absence of TBA (control, Figure 2A) and after preincubation for 330 s with 20 mM TBA while holding at −100 mV. Subsequently, 0.3 Hz pulse trains were applied until steady state block was reached. The ratio between tail current amplitude in the presence of TBA and tail current amplitude in the control solution was taken as a measure of block. Channel block developed in a "use-dependent" manner. The first current after the 330 s equilibrium in TBA displayed a pronounced decay during the 300 ms prepulse to +20 mV (Figure 2A, p1) with a significant tail current inhibition. Prepulse and tail currents were further inhibited during the 0.3 Hz pulse train. The steady

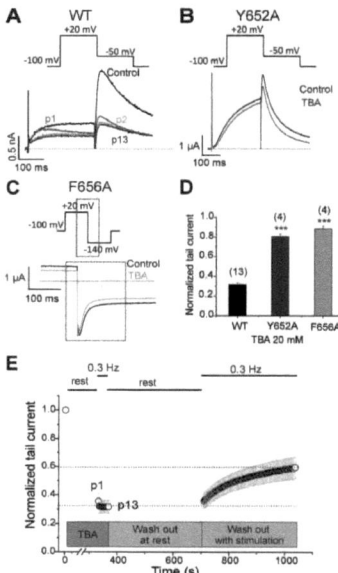

Figure 2. WT and mutant hERG channel inhibition by TBA. A) Superimposed current traces in the absence of TBA (control, black) and after a 330 s preincubation period in 20 mM TBA (p1, magenta). Steady state block occurred within the first 2 pulses (p2, orange and p13, brown). The voltage protocol is shown on top of the current traces. B and C) Representative current traces and corresponding voltage protocols for current measurements of mutants Y652A and F656A in the absence (Control, black trace) and presence of 20 mM TBA (blue and orange trace, respectively). Tail currents of F656A were recorded at −140 mV. D) Normalized peak tail currents of WT, Y652A, and F656A channels after steady state block by 20 mM TBA ($n = 4$–13, error bars, ± SEM; unpaired *t*-test, $P < 0.0001$). E) State dependent recovery of hERG channels from TBA block. Peak tail currents were normalized to control currents and plotted against time. After 330 s incubation of WT hERG channels with 20 mM TBA, steady state block was reached within the first 2 pulses of a 0.3 Hz pulse train (p1–p13). During the following 330 s wash period, channels were kept closed at resting potential of −100 mV. Recovery from block at rest was probed by subsequent pulsing at 0.3 Hz.

state block was achieved rapidly within the first 2 pulses (Figure 2A; p2, p13). The development of block during channel activation at +20 mV suggests that TBA blocks hERG channels in an open channel conformation. Twenty mM TBA blocked hERG channels by 68.3 ± 2.0% (Figure 2D).

To identify amino acids essential for TBA block, we performed alanine mutation studies on Y652 and F656 which have been shown to play a key role for binding of different chemical entities.[16] The WT channel voltage protocol was utilized for Y652A, while tail currents were measured at −140 mV for F656A as reported by Witchel et al.[15] Y652A and F656A significantly reduced channel inhibition to 19.6 ± 2.7% and 11.7 ± 3.2%, respectively (Figure 2B–D). This is in line with data shown in Figure 2A suggesting that TBA accesses a binding site inside the cavity comprising of the two prominent aromatic residues.

Next, we probed if TBA is trapped inside the hERG cavity as suggested for KcsA.[30,37,38] The hallmark of drug trapping is an ultraslow recovery or lack of recovery at rest.[18,22−24] Recovery of hERG from TBA block during a 0.3 Hz train was monitored after a 330 s period at holding potential of −100 mV where the channels are in a closed resting state. During this rest period, the oocytes were perfused with TBA-free solution (Figure 2E). The first current amplitudes after this rest period recovered from block only by 5 ± 1.1% indicating that TBA is trapped in the closed channel conformation. Subsequent frequent opening of the channel during continuous stimulation at 0.3 Hz induced monoexponential recovery from TBA block to 41.3 ± 8.3% in 330 s (Figure 2E) suggesting that trapped TBA leaves the channel during activation when the channels are in an open conformation.

Structural Insights of TBA Block in hERG. To further explore TBA hERG channel interactions, we docked TBA into the hERG cavity using the program FlexX and the open hERG homology model.[40] In the docking simulations, the tilted orientation is favored over the xy-plane orientation thereby maximizing hydrophobic interactions of TBA with Y652 and F656 (Figure 3A and B). In a subsequent 100 ns MD

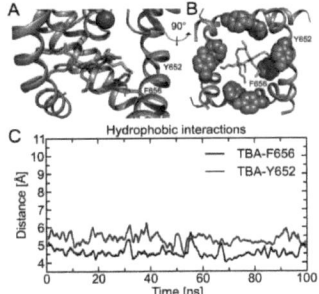

Figure 3. Orientation of TBA in hERG and distances between TBA side chains and Y652 and F656. A) TBA in the xy-plane (pink sticks) and tilted (magenta sticks) orientation in side view. For clarity, only three SUs are shown in gray. The K$^+$ ion in the SF is represented as a violet sphere. Y652 and F656 are shown as green sticks. B) Top view of TBA in the xy-plane (pink sticks) and tilted (magenta sticks) orientation. Y652 and F656 in the cavity facing state are shown as green spheres. C) Distances between TBA side chains and Y652 and F656. The distance between the three outermost carbon atoms of each butyl side chain and the aromatic rings was calculated. The closest distance at each time step is plotted.

simulation starting from the best ranked docking pose, TBA sampled both planar and tilted orientations equally, closely interacting with the two aromatic amino acids Y652 and F656. In Figure 3C, the shortest distances between a TBA side chain and the aromatic rings of Y652 and F656 are plotted over time. Distances between 4 and 6 Å indicate that TBA side chains favorably interact with the aromatic amino acids. Further, the central ammonium group of TBA is positioned in the center of the pore, near the K$^+$ binding site, suggesting that the helix dipole charges from the pore helices contribute to binding.

Measured distances between the quaternary nitrogen of TBA and Y652 and F656 were always above 5.5 Å in our simulation. At such distances the potential energies of cation-pi interactions become insignificant.[64] Further, interactions between the aromatic rings and the H atoms on the carbons adjacent to the protonated nitrogen, as described by Imai et al.,[74] were above 4.5 Å. Therefore, it is unlikely that cation-pi stacking contributes to binding. Throughout the simulations TBA remained in the energetically favorable D$_{2d}$ conformation.

Dynamics of TBA Trapping in hERG. To investigate possible conformational changes during hERG gating, we performed 20 ED simulations in the absence and presence of TBA. The apo simulations revealed that the conformation of Y652 remains in the cavity facing orientation, independent of channel state (Figure 4A). In contrast the rotameric state of the second aromatic amino acid forming the binding site, F656 changed dramatically during gate closure. While preferably in the cavity facing conformation in the open channel state, F656 switched rapidly to the cavity lining conformation in the apo simulations, reaching equal distribution between the two states after 8 ns (Figure 4B). Interestingly, a third, rare rotameric state of F656 was observed during gating. F656 can rotate to a state orthogonally to the S6 helix since two adjacent S6 helices approach each other during closing and therefore decrease the space in this region. These findings show that Y652 remains rigid during gating while F656 undergoes gating specific rotameric changes.

Next, we probed the influence of TBA on gating dependent rearrangements of binding residues. We observed no significant differences for Y652 (Figure 4C). However, TBA clearly influenced the dynamics of F656 during trapping. Figure 4D illustrates that TBA slows the transition from the cavity facing to the cavity lining conformation. In simulations without the blocker, equal distribution was reached after 8 ns, while in the presence of TBA, 15 ns were necessary. This suggests that the trapped TBA stabilizes the F656 conformation in cavity facing orientation.

As observed in KcsA, TBA remains in the D$_{2d}$ conformation with rare observations of the S$_4$ state during hERG channel closure (supplemental Figure S2). Due to the larger cavity, TBA can adopt the tilted as well as the xy-plane orientation (Figure 4E and F).

State Dependence of TBA Dissociation. While it was shown that dissociation of trapped compounds is explicitly linked to an open gate, the extent of gate opening necessary for dissociation remains elusive. Crystal structures of KcsA in closed, intermediate, and open states[39,65] and a cocrystallized TBA[37,38] provide an ideal set of structural information to probe state dependent dissociation. The closed channel with TBA, crystallized by Yohannan et al.,[38] served as a starting point. TBA was placed to the same binding site in the intermediate and open channel state,[39] and all systems were subject to 20 ns free MD simulation to allow equilibration of TBA orientation.

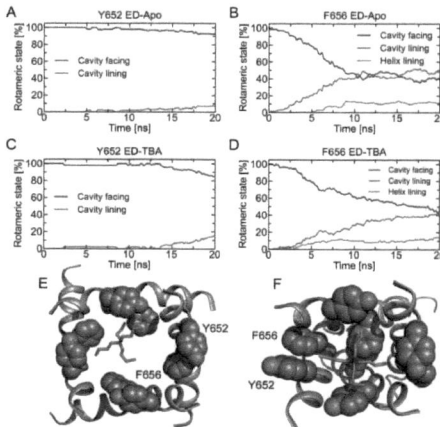

Figure 4. Rotameric states of Y652 and F656 during hERG gate closure. Conformational changes of Y652 (A and C) and F656 (B and D) in the absence (A and B) and presence (C and D) of TBA during gating. Cavity facing (blue line) and cavity lining (green line) states identified for KcsA also apply for hERG. For F656, an additional helix lining state ($\chi 1$ angle in the range of 80°, red line) was observed. E and F) Rotameric states of Y652 and F656 as spheres at the beginning and end of gate closure in top view, respectively. TBA is represented as sticks and colored magenta and pink in the tilted and xy-plane orientation, respectively. Y652 and F656 are colored according to their rotameric states.

TBA remained in the xy-plane orientation as seen in the crystal structures.[37,38] Subsequently, force probe MD simulations were used to pull the compound along the channel axis through the gate region. Free energy calculations of the dissociation pathway by umbrella sampling provide qualitative insights into the probability of TBA dissociation from specific channel states. In the closed KcsA channel, TBA is tightly packed in the cavity formed by F103 and I100 as well as by T107 where the pore becomes constricted (see Figure 5B and C). The small energy well in the cavity and the large energy increase along the z-axis indicate that the cavity space in the closed state is very limited (Figure 5A). Squeezing TBA through the closed gate (Figure 5D−G) led to a total energy barrier of 50 kcal/mol rendering spontaneous dissociation through the closed gate very unlikely. Even at the membrane solvent interface region the barrier remains high, due to unfavorable interactions of TBA with the pH sensing residues of KcsA (Figure 5H). During the force probe simulation, TBA adopted and remained in the S_4 conformation. This suggests that part of the applied energy, contributing to the energy barrier, is used to change the TBA conformation. In addition, up to 1 kcal/mol might be stored in the S_4 conformation.

Next, TBA was pulled through the intermediate channel gate (gate diameter of 8 Å). The opening of the gate increased the cavity size leading to a broader energy well of TBA in the binding site. However, an energy barrier of 40 kcal/mol was calculated for TBA dissociation (Figure 6A−G). Again, TBA adopted the S_4 conformation during dissociation and stayed in that conformation. Similar to dissociation from the closed state the barrier remains high, when reaching the membrane solvent interface region. As can be seen in Figure 6G, in the region of 4.2 to 4.6 nm on the z-axis, TBA passes the pH sensing residues of KcsA.

In the open KcsA channel, the cavity (gate diameter of 14 Å) is directly accessible to the solvent. The force probe MD simulations revealed that TBA, due to its hydrophobic nature, moves along the TM2 and TM1 helices before it gets fully hydrated in the intracellular compartment. Free energy calculations showed that the open gate does not cause an energy barrier for TBA dissociation (Figure 7A). The increase in energy during dissociation is mainly caused by TBA leaving its high affinity binding site between 2 and 2.5 nm on the z-axis (Figure 7A, B−D). A total energy difference of 8 kcal/mol between the bound and the solvated TBA was measured. Throughout the dissociation simulation, TBA remained in the D_{2d} conformation further indicating that the gate is wide enough to allow dissociation of a planar TBA which has a diameter of 12 Å (Figure 7B−F).

TBA Dissociation from the Open hERG. To explore dissociation of TBA from the hERG channel we applied force probe simulations and umbrella sampling, as described above for KcsA. TBA dissociation was only probed from the fully open hERG state since free energy calculations on KcsA suggest that dissociation solely occurs from an open channel state.

For hERG dissociation, a total energy difference of 6 kcal/mol was measured (Figure 8A−C). At position 2.2 to 2.4 nm on the z-axis, an interacting F656 switched from the cavity facing to the cavity lining state, thereby allowing easier passage

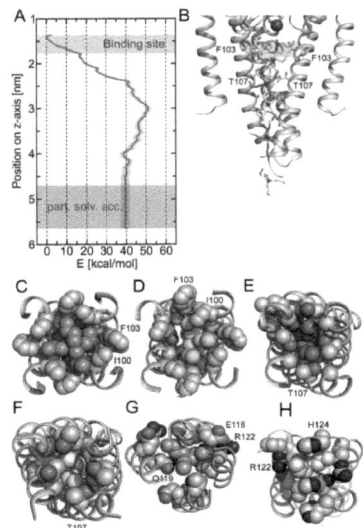

Figure 5. A) Free energy profile of TBA dissociation from the closed KcsA channel. The blue and gray shades depict the partially solvent accessible membrane water interface (part. solv. acc.), respectively. Statistical error is shown as green shade. B) TBA structures, represented as sticks, show the dissociation pathway through the gate. For clarity, only three SUs are shown in green. Interacting amino acids are depicted as sticks. The K$^+$ ion in the SF is represented as a violet sphere. TBA structures in panel B and C–H correspond to positions 1.45 nm (pink, C), 1.84 nm (cyan, D), 2.45 nm (magenta, E), 3.0 nm (yellow, F), 4.0 nm (light pink, G), and 5.3 nm (gray, H) on the z-axis. C–G) Top view of TBA and interacting amino acids during dissociation. H) Bottom view of TBA with interacting charged residues from the pH sensor. TBA is colored according to its position, and amino acids are shown as green spheres with one SU labeled.

of TBA through the channel gate and causing an energy plateau phase. The subsequent energy barrier from 2.5 to 3 nm is caused by the first exposure of TBA to the solvent after the F656 passage (Figure 8D and E). At 3 nm on the z-axis (Figure 8E), TBA packs to two adjacent TM2 helices before it gets fully hydrated shown by an energy increase to 6 kcal/mol. Throughout the dissociation simulation, TBA remained in the D_{2d} conformation.

■ DISCUSSION

In this study we addressed two important questions concerning drug trapping in K$^+$ channels. First, do trapped drugs influence gating? Second, is a fully open gate required for drug dissociation? To answer the first question, we performed MD

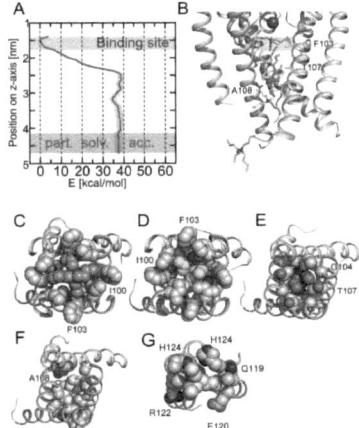

Figure 6. A) Free energy profile of TBA dissociation from the intermediate KcsA channel. The blue and gray shades depict the high affinity binding site and the partially solvent accessible membrane water interface (part. solv. acc.), respectively. Statistical error is shown as green shade. B) TBA structures, represented as sticks, show the dissociation pathway through the gate. For clarity, only three SUs are shown in green. Interacting amino acids are depicted as green sticks. The K$^+$ ion in the SF is represented as a violet sphere. TBA structures in panels B and C–F correspond to positions 1.5 nm (pink, C), 2.1 nm (cyan, D), 2.5 nm (magenta, E), 3.2 nm (yellow, F), and 4.4 nm on the z-axis (gray, G). C–F) Top view of TBA and interacting amino acids during dissociation. TBA is colored according to its position, and amino acids are shown as green spheres with one SU labeled. G) Bottom view of TBA and interacting charged residues from the pH sensor.

simulations with the prototypical K$^+$ channel KcsA and a model of the hERG K$^+$ channel. Closing ED simulations with TBA in KcsA and hERG revealed that the drug influences structural rearrangements in hERG but not in KcsA. This is in agreement with crystal structures of TBA trapped in KcsA by Faraldo-Gómez et al.[37] and Yohannan et al.[38] indicating a negligible effect of drug block on the channel structure.

In hERG, experimental characterization of TBA block was lacking so far. Thus, we first investigated the mechanism of TBA hERG interactions in detail. Open/inactivated channel dependence of block of TBA was indicated by fast current decrease upon channel activation (Figure 2A). This finding is in line with the well described state dependence of drug block in hERG.[66,67] Alanine mutations of Y652 and F656 revealed that both residues significantly reduce the potency of TBA (Figure 2D) indicating that this compound binds to the cavity as has been shown for a multitude of other hERG blockers.[7−10,12−14,18−20]

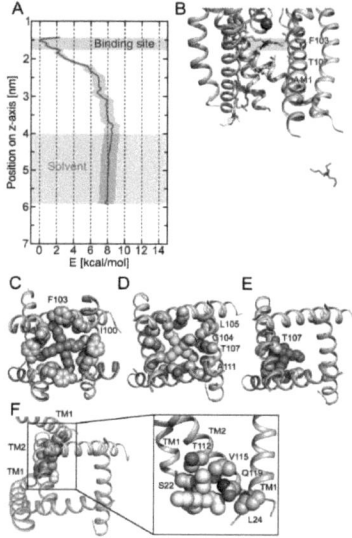
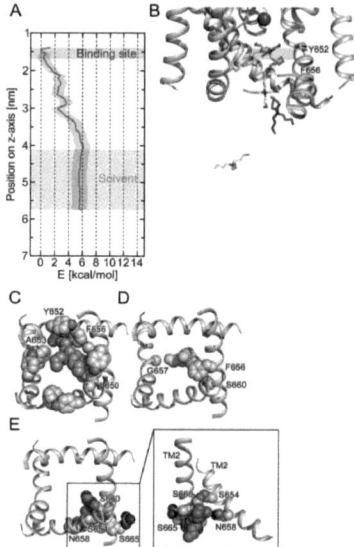

Figure 7. A) Free energy profile of TBA dissociation from the fully open KcsA channel state. The blue and beige shades depict the high affinity binding site and the solvent compartment, respectively. Statistical error is shown as green shade. B) TBA structures, represented as sticks, show the dissociation pathway through the gate. For clarity, only three SUs are shown in green. Interacting amino acids are depicted as green sticks. TBA structures in panels B and C–F correspond to positions 1.5 nm (pink, C), 2.3 nm (cyan, D), 2.9 nm (magenta, E), 3.6 nm (yellow, F), and 5.2 nm on the z-axis. C–E) Top view of TBA and interacting amino acids during dissociation. F) Top and side view of TBA and interacting amino acids on TM1 and TM2 helices.

Figure 8. A) Free energy profile of TBA dissociation from the open hERG channel. The blue and beige shades depict the binding site and the solvent compartment, respectively. Statistical error is shown as green shade. B) TBA structures, represented as sticks, show the dissociation pathway through the gate. For clarity, only three SUs are shown in green. The binding determinants Y652 and F656 are depicted as green sticks. TBA structures in panels B and C–E correspond to positions 1.58 nm (pink, C), 2.5 nm (cyan, D), 3 nm (magenta, E), and 4.5 nm (yellow) on the z-axis. C–D) Top view of TBA and interacting amino acids during dissociation. E) Top and side view of TBA and interacting amino acids on two adjacent TM2 helices.

TBA shares common features with trapped drugs. Block occurs from the intracellular side of the channel and requires prior channel opening. Further, the drug cannot be washed out during 330 s at resting state and repetitive stimulation during wash-out induces rapid recovery from block, suggesting that TBA is trapped in the hERG cavity (Figure 2E).

In agreement with experimental results, docking and MD simulations support the importance of Y652 and F656 for binding. According to our modeling studies, these interactions are primarily of hydrophobic nature. Cation-pi interactions were not observed; the quaternary nitrogen and aromatic rings of Y652 and F656 were too distant[64] throughout the simulation. This fits to data by Xia et al. showing that higher exposure of the quaternary nitrogen and a decrease of hydrophobicity by shorter alkyl chains leads to reduced potency of QA compounds.[68] Another important component

to TBA binding in hERG results from focused helix dipole charges from the pore helices. This is in agreement with a recent study by Dempsey et al.,[75] predicting similar interactions for TBA interactions in KcsA.

The hERG channel closure simulations highlight the important role of F656 during channel closure. Fast transitions of F656 from cavity facing to cavity lining conformation were observed (Figure 4B). This finding is in agreement with a recent MD simulation study revealing innate flexibility of the F656 side chain.[69] Further, it is supported by a mutagenesis study by Fernandez et al.,[16] suggesting an important role of this residue for normal deactivation. Their work cleary showed that replacement of the bulky side chain by smaller hydrophobic amino acids leads to faster channel closure. Therefore, it is conceivable that the bulkiness at this position contributes to the slow deactivation kinetics in hERG.

Remarkably, in the presence of TBA, F656 structural rearrangements are significantly perturbed during gating (Figure 4D), indicating that binding of TBA stabilizes the aromatic side chains in the cavity facing conformation. This finding is of particular interest since most hERG blockers bind to F656.[7−10,12−14,18−20] Of note, a study by Chen et al.[76] demonstrated that inactivation leads to gating associated reorientations of Y652 and F656. To what extent the observed conformational changes in our simulation might contribute to inactivation needs to be addressed in future studies.

Interestingly, it has been suggested previously that F656 might act as physical barrier for drug dissociation of certain compounds.[70] It is tempting to speculate that specific interactions with F656 determine if a compound is trapped or not. Based on our simulations, we propose that trapped compounds might stabilize the cavity facing state of F656, presenting a barrier for drug dissociation. Increased sampling of the cavity lining conformation might facilitate drug dissociation prior to complete channel closure. This hypothetical mechanism is illustrated in schematic Figure 9. One way to test the

Figure 9. Schematic figure of the importance of the F656 conformation on trapping.

validity of this scenario would be the use of F656 mutants as described by Fernandez et al.,[16] which still exhibit reasonable affinity for blockers while introducing different amino acid size and deactivation kinetics. This approach will be the subject of further studies.

In hERG, not all blockers demonstrate slow recovery from block[23,71,72] but appear to quickly dissociate even when the channels are held at resting state. This might be explained by our findings, demonstrating the important role of F656 during gating. F656 might function as a second gate and prevent dissociation of trapped drugs via mutual interference.

The second important question that was addressed in this study concerns the extent of gate opening necessary for dissociation. Force probe MD simulations and energy calculations on 3 different KcsA crystal structures with pore diameters of up to 14 Å (fully open) revealed that dissociation is only possible when the gate is fully open. Similarly, no energy barrier was found for TBA dissociation from the open hERG state model. In both channels, TBA moves along the cavity wall maximizing hydrophobic contacts to the protein during dissociation. The exit scenario from closed and intermediate KcsA channel states is predicted to trigger conformational changes of TBA to the S_4 conformation. Despite this more compressed structure (diameter of 8 Å), large energy barriers render dissociation from these states highly unlikely. Since except for F103, only small residues line the cavity in KcsA, the

observed high energy barriers (40−50 kcal/mol) result primarily from the conformational state, defined by the backbone coordinates. Although high energy barriers occurred during dissociation from these channel states, our force probe simulations did not cause changes in the protein secondary structure. Taken together, we propose that compounds cannot dissociate from closed or intermediate states but require an open helix bundle crossing gate. This is supported by earlier findings from our lab.[73]

■ ASSOCIATED CONTENT

◯ Supporting Information

Distribution of dihedral angles of TBA monitored during closing ED simulations in KcsA and hERG and histograms of the umbrella sampling windows. This material is available free of charge via the Internet at http://pubs.acs.org.

■ AUTHOR INFORMATION

Corresponding Author
*E-mail: anna.stary@univie.ac.at.

Author Contributions
Conceived and designed the experiments: T.L., P.S., E.T., S.H., A.S.W. Performed the experiments: T.L., P.S. Analyzed the data: T.L., P.S., E.T., S.H., A.S.W. Wrote the paper: T.L., P.S., E.T., S.H., A.S.W. All authors have given approval to the final version of the manuscript.

Funding
This work was supported by the Austrian Science Fund (FWF; Grants P22395 and W1232; http://www.fwf.ac.at). Tobias Linder was supported by a research fellowship from the University of Vienna and an EMBO short-term fellowship. Anna Stary-Weinzinger is supported by the Johanna Mahlke, geb. Obermann Stiftung.

Notes
The authors declare no competing financial interest.

■ ACKNOWLEDGMENTS

The computational results presented have been achieved using the Vienna Scientific Cluster (VSC).

■ ABBREVIATIONS

ED, essential dynamics; hERG, human ether-à-go-go related gene; LQTS, long QT syndrome; SF, selectivity filter; TBA, tetrabutylammonium; TEA, tetraethylammonium; QA, quaternary ammonium

■ REFERENCES

(1) Bagal, S. K.; Brown, A. D.; Cox, P. J.; Omoto, K.; Owen, R. M.; Pryde, D. C.; Sidders, B.; Skerratt, S. E.; Stevens, E. B.; Storer, R. I.; Swain, N. A. Ion Channels as Therapeutic Targets: A Drug Discovery Perspective. *J. Med. Chem.* **2013**, *56*, 593−624.
(2) Fermini, B.; Fossa, A. A. The Impact of Drug-Induced QT Interval Prolongation on Drug Discovery and Development. *Nat. Rev. Drug Discovery* **2003**, *2*, 439−447.
(3) Chiang, C. E.; Roden, D. M. The Long QT Syndromes: Genetic Basis and Clinical Implications. *J. Am. Coll. Cardiol.* **2000**, *36*, 1−12.
(4) Keating, M. T.; Sanguinetti, M. C. Molecular and Cellular Mechanisms of Cardiac Arrhythmias. *Cell* **2001**, *104*, 569−580.
(5) Jehle, J.; Schweizer, P. A.; Katus, H. A.; Thomas, D. Novel Roles for hERG K(+) Channels in Cell Proliferation and Apoptosis. *Cell Death Dis.* **2011**, *2*, e193.
(6) Pier, D. M.; Shehatou, G. S.; Giblett, S.; Pullar, C. E.; Tresize, D. J.; Pritchard, C. A.; Challiss, J.; Mitcheson, J. S. Long-Term Channel

Block Is Required to Inhibit Cellular Transformation by Human Ether-a-Go-Go-Related Gene (Herg1) Potassium Channels. *Mol. Pharmacol.* **2014**, *86*, 211–21.

(7) Mitcheson, J. S.; Chen, J.; Lin, M.; Culberson, C.; Sanguinetti, M. C. A Structural Basis for Drug-Induced Long QT Syndrome. *Proc. Natl. Acad. Sci. U. S. A.* **2000**, *97*, 12329–12333.

(8) Lees-Miller, J. P.; Duan, Y.; Teng, G. Q.; Duff, H. J. Molecular Determinant of High-Affinity Dofetilide Binding to HERG1 Expressed in Xenopus Oocytes: Involvement of S6 Sites. *Mol. Pharmacol.* **2000**, *57*, 367–374.

(9) Kamiya, K.; Mitcheson, J. S.; Yasui, K.; Kodama, I.; Sanguinetti, M. C. Open Channel Block of HERG K(+) Channels by Vesnarinone. *Mol. Pharmacol.* **2001**, *60*, 244–253.

(10) Sánchez-Chapula, J. A.; Navarro-Polanco, R. A.; Culberson, C.; Chen, J.; Sanguinetti, M. C. Molecular Determinants of Voltage-Dependent Human Ether-a-Go-Go Related Gene (HERG) K+ Channel Block. *J. Biol. Chem.* **2002**, *277*, 23587–23595.

(11) Sánchez-Chapula, J. A.; Ferrer, T.; Navarro-Polanco, R. A.; Sanguinetti, M. C. Voltage-Dependent Profile of Human Ether-A-Go-Go Related Gene Channel Block Is Influenced by a Single Residue in the S6 Transmembrane Domain. *Mol. Pharmacol.* **2003**, *63*, 1051–1058.

(12) Sánchez-Chapula, J. A.; Navarro-Polanco, R. A.; Sanguinetti, M. C. Block of Wild-Type and Inactivation-Deficient Human Ether-a-Go-Go-Related Gene K+ Channels by Halofantrine. *Naunyn-Schmiedeberg's Arch. Pharmacol.* **2004**, *370*, 484–491.

(13) Ridley, J. M.; Dooley, P. C.; Milnes, J. T.; Witchel, H. J.; Hancox, J. C. Lidoflazine Is a High Affinity Blocker of the HERG K(+)channel. *J. Mol. Cell. Cardiol.* **2004**, *36*, 701–705.

(14) Perry, M.; de Groot, M. J.; Helliwell, R.; Leishman, D.; Tristani-Firouzi, M.; Sanguinetti, M. C.; Mitcheson, J. Structural Determinants of HERG Channel Block by Clofilium and Ibutilide. *Mol. Pharmacol.* **2004**, *66*, 240–249.

(15) Witchel, H. J.; Dempsey, C. E.; Sessions, R. B.; Perry, M.; Milnes, J. T.; Hancox, J. C.; Mitcheson, J. S. The Low-Potency, Voltage-Dependent HERG Blocker Propafenone - Molecular Determinants and Drug Trapping. *Mol. Pharmacol.* **2004**, *66*, 1201–1212.

(16) Fernandez, D.; Ghanta, A.; Kauffman, G. W.; Sanguinetti, M. C. Physicochemical Features of the HERG Channel Drug Binding Site. *J. Biol. Chem.* **2004**, *279*, 10120–10127.

(17) Sanguinetti, M. C.; Mitcheson, J. S. Predicting Drug-hERG Channel Interactions That Cause Acquired Long QT Syndrome. *Trends Pharmacol. Sci.* **2005**, *26*, 119–124.

(18) Kamiya, K.; Niwa, R.; Mitcheson, J. S.; Sanguinetti, M. C. Molecular Determinants of HERG Channel Block. *Mol. Pharmacol.* **2006**, *69*, 1709–1716.

(19) Guo, J.; Gang, H.; Zhang, S. Molecular Determinants of Cocaine Block of Human Ether-Á-Go-Go-Related Gene Potassium Channels. *J. Pharmacol. Exp. Ther.* **2006**, *317*, 865–874.

(20) Kamiya, K.; Niwa, R.; Morishima, M.; Honjo, H.; Sanguinetti, M. C. Molecular Determinants of hERG Channel Block by Terfenadine and Cisapride. *J. Pharmacol.* **2008**, *108*, 301–307.

(21) Carmeliet, E. Voltage- and Time-Dependent Block of the Delayed K+ Current in Cardiac Myocytes by Dofetilide. *J. Pharmacol. Exp. Ther.* **1992**, *262*, 809–817.

(22) Mitcheson, J. S.; Chen, J.; Sanguinetti, M. C. Trapping of a Methanesulfonanilide by Closure of the HERG Potassium Channel Activation Gate. *J. Gen. Physiol.* **2000**, *115*, 229–240.

(23) Stork, D.; Timin, E. N.; Berjukow, S.; Huber, C.; Hohaus, A.; Auer, M.; Hering, S. State Dependent Dissociation of HERG Channel Inhibitors. *Br. J. Pharmacol.* **2007**, *151*, 1368–1376.

(24) Windisch, A.; Timin, E.; Schwarz, T.; Stork-Riedler, D.; Erker, T.; Ecker, G.; Hering, S. Trapping and Dissociation of Propafenone Derivatives in HERG Channels. *Br. J. Pharmacol.* **2011**, *162*, 1542–1552.

(25) Di Veroli, G. Y.; Davies, M. R.; Zhang, H.; Abi-Gerges, N.; Boyett, M. R. hERG Inhibitors With Similar Potency But Different Binding Kinetics Do Not Pose the Same Proarrhythmic Risk: Implications for Drug Safety Assessment. *J. Cardiovasc. Electrophysiol.* **2013**, *25*, 197–207.

(26) Armstrong, C. M. Interaction of Tetraethylammonium Ion Derivatives with the Potassium Channels of Giant Axons. *J. Gen. Physiol.* **1971**, *58*, 413–437.

(27) Armstrong, C. M.; Hille, B. The Inner Quaternary Ammonium Ion Receptor in Potassium Channels of the Node of Ranvier. *J. Gen. Physiol.* **1972**, *59*, 388–400.

(28) MacKinnon, R.; Yellen, G. Mutations Affecting TEA Blockade and Ion Permeation in Voltage-Activated K+ Channels. *Science* **1990**, *250*, 276–279.

(29) Luzhkov, V. B.; Aqvist, J. Mechanisms of Tetraethylammonium Ion Block in the KcsA Potassium Channel. *FEBS Lett.* **2001**, *495*, 191–196.

(30) Zhou, M.; Morais-Cabral, J. H.; Mann, S.; MacKinnon, R. Potassium Channel Receptor Site for the Inactivation Gate and Quaternary Amine Inhibitors. *Nature* **2001**, *411*, 657–661.

(31) Holmgren, M.; Smith, P. L.; Yellen, G. Trapping of Organic Blockers by Closing of Voltage-Dependent K+ Channels: Evidence for a Trap Door Mechanism of Activation Gating. *J. Gen. Physiol.* **1997**, *109*, 527–535.

(32) Posson, D. J.; McCoy, J. G.; Nimigean, C. M. The Voltage-Dependent Gate in MthK Potassium Channels Is Located at the Selectivity Filter. *Nat. Struct. Mol. Biol.* **2013**, *20*, 159–166.

(33) Yellen, G.; Jurman, M. E.; Abramson, T.; MacKinnon, R. Mutations Affecting Internal TEA Blockade Identify the Probable Pore-Forming Region of a K+ Channel. *Science (80-.)* **1991**, *251*, 939–942.

(34) Choi, K. L.; Mossman, C.; Aubé, J.; Yellen, G. The Internal Quaternary Ammonium Receptor Site of Shaker Potassium Channels. *Neuron* **1993**, *10*, 533–541.

(35) Crouzy, S.; Bernèche, S.; Roux, B. Extracellular Blockade of K(+) Channels by TEA: Results from Molecular Dynamics Simulations of the KcsA Channel. *J. Gen. Physiol.* **2001**, *118*, 207–218.

(36) Lenaeus, M. J.; Vamvouka, M.; Focia, P. J.; Gross, A. Structural Basis of TEA Blockade in a Model Potassium Channel. *Nat. Struct. Mol. Biol.* **2005**, *12*, 454–459.

(37) Faraldo-Gómez, J. D.; Kutluay, E.; Jogini, V.; Zhao, Y.; Heginbotham, L.; Roux, B. Mechanism of Intracellular Block of the KcsA K+ Channel by Tetrabutylammonium: Insights from X-Ray Crystallography, Electrophysiology and Replica-Exchange Molecular Dynamics Simulations. *J. Mol. Biol.* **2007**, *365*, 649–662.

(38) Yohannan, S.; Hu, Y.; Zhou, Y. Crystallographic Study of the Tetrabutylammonium Block to the KcsA K+ Channel. *J. Mol. Biol.* **2007**, *366*, 806–814.

(39) Cuello, L. G.; Jogini, V.; Cortes, D. M.; Perozo, E. Structural Mechanism of C-Type Inactivation in K(+) Channels. *Nature* **2010**, *466*, 203–208.

(40) Stary, A.; Wacker, S. J.; Boukharta, L.; Zachariae, U.; Karimi-Nejad, Y.; Aqvist, J.; Vriend, G.; de Groot, B. L. Toward a Consensus Model of the HERG Potassium Channel. *ChemMedChem* **2010**, *5*, 455–467.

(41) Wang, J.; Wolf, R. M.; Caldwell, J. W.; Kollman, P. A.; Case, D. A. Development and Testing of a General Amber Force Field. *J. Comput. Chem.* **2004**, *25*, 1157–1174.

(42) Frisch, M. J.; Trucks, G. W.; Schlegel, H. B.; Scuseria, G. E.; Robb, M. A.; Cheeseman, J. R.; Scalmani, G.; Barone, V.; Mennucci, B.; Petersson, G. A.; Nakatsuji, H.; Caricato, M.; Li, X.; Hratchian, H. P.; Izmaylov, A. F.; Bloino, J.; Zheng, G.; Sonnenberg, J. L.; Hada, M.; Ehara, M.; Toyota, K.; Fukuda, R.; Hasegawa, J.; Ishida, M.; Nakajima, T.; Honda, Y.; Kitao, O.; Nakai, H.; Vreven, T.; Montgomery, J. A., Jr.; Peralta, J. E.; Ogliaro, F.; Bearpark, M.; Heyd, J. J.; Brothers, E.; Kudin, K. N.; Staroverov, V. N.; Kobayashi, R.; Normand, J.; Raghavachari, K.; Rendell, A.; Burant, J. C.; Iyengar, S. S.; Tomasi, J.; Cossi, M.; Rega, N.; Millam, J. M.; Klene, M.; Knox, J. E.; Cross, J. B.; Bakken, V.; Adamo, C.; Jaramillo, J.; Gomperts, R.; Stratmann, R. E.; Yazyev, O.; Austin, A. J.; Cammi, R.; Pomelli, C.; Ochterski, J. W.; Martin, R. L.; Morokuma, K.; Zakrzewski, V. G.; Voth, G. A.; Salvador, P.; Dannenberg, J. J.; Dapprich, S.; Daniels, A. D.; Farkas, Ö.;

(43) Wang, J.; Wang, W.; Kollman, P. A.; Case, D. A. Automatic Atom Type and Bond Type Perception in Molecular Mechanical Calculations. *J. Mol. Graphics Modell.* **2006**, *25*, 247−260.
(44) Case, D. A.; Darden, T. A.; Cheatham, T. E.; Simmerling, C. L.; Wang, J.; Duke, R. E.; Luo, R.; Crowley, M.; Walker, R. C.; Zhang, W.; Merz, K. M.; Wang, B.; Hayik, S.; Roitberg, A.; Seabra, G.; Kolossváry, I.; Wong, K. F.; Paesani, F.; Vanicek, J.; Wu, X.; Brozell, S. R.; Steinbrecher, T.; Gohlke, H.; Yang, L.; Tan, C.; Mongan, J.; Hornak, V.; Cui, G.; Mathews, D. H.; Seetin, M. G.; Sagui, C.; Babin, V.; Kollman, P. A. *Amber 11*; 2010.
(45) Wolf, M. G.; Hoefling, M.; Aponte-Santamaría, C.; Grubmüller, H.; Groenhof, G. G_membed: Efficient Insertion of a Membrane Protein into an Equilibrated Lipid Bilayer with Minimal Perturbation. *J. Comput. Chem.* **2010**, *31*, 2169−2174.
(46) Aqvist, J.; Luzhkov, V. Ion Permeation Mechanism of the Potassium Channel. *Nature* **2000**, *404*, 881−884.
(47) Hess, B.; Kutzner, C.; van der Spoel, D.; Lindahl, E. GROMACS 4: Algorithms for Highly Efficient, Load-Balanced, and Scalable Molecular Simulation. *J. Chem. Theory Comput.* **2008**, *4*, 435−447.
(48) Hornak, V.; Abel, R.; Okur, A.; Strockbine, B.; Roitberg, A.; Simmerling, C. Comparison of Multiple Amber Force Fields and Development of Improved Protein Backbone Parameters. *Proteins* **2006**, *65*, 712−725.
(49) Jorgensen, W. L.; Chandrasekhar, J.; Madura, J. D.; Impey, R. W.; Klein, M. L. Comparison of Simple Potential Functions for Simulating Liquid Water. *J. Chem. Phys.* **1983**, *79*, 926.
(50) Berger, O.; Edholm, O.; Jähnig, F. Molecular Dynamics Simulations of a Fluid Bilayer of Dipalmitoylphosphatidylcholine at Full Hydration, Constant Pressure, and Constant Temperature. *Biophys. J.* **1997**, *72*, 2002−2013.
(51) Joung, I. S.; Cheatham, T. E. Determination of Alkali and Halide Monovalent Ion Parameters for Use in Explicitly Solvated Biomolecular Simulations. *J. Phys. Chem. B* **2008**, *112*, 9020−9041.
(52) Darden, T.; York, D.; Pedersen, L. Particle Mesh Ewald: An N log(N) Method for Ewald Sums in Large Systems. *J. Chem. Phys.* **1993**, *98*, 10089.
(53) Hess, B.; Bekker, H.; Berendsen, H. J. C.; Fraaije, J. G. E. M. LINCS: A Linear Constraint Solver for Molecular Simulations. *J. Comput. Chem.* **1997**, *18*, 1463−1472.
(54) Nosé, S. A Unified Formulation of the Constant Temperature Molecular Dynamics Methods. *J. Chem. Phys.* **1984**, *81*, 511.
(55) Hoover, W. Canonical Dynamics: Equilibrium Phase-Space Distributions. *Phys. Rev. A* **1985**, *31*, 1695−1697.
(56) Parrinello, M.; Rahman, A. Polymorphic Transitions in Single Crystals: A New Molecular Dynamics Method. *J. Appl. Phys.* **1981**, *52*, 7182.
(57) Linder, T.; de Groot, B. L.; Stary-Weinzinger, A. Probing the Energy Landscape of Activation Gating of the Bacterial Potassium Channel KcsA. *PLoS Comput. Biol.* **2013**, *9*, e1003058.
(58) Hub, J. S.; de Groot, B. L.; van der Spoel, D. g_wham—A Free Weighted Histogram Analysis Implementation Including Robust Error and Autocorrelation Estimates. *J. Chem. Theory Comput.* **2010**, *6*, 3713−3720.
(59) Sanguinetti, M. C.; Jiang, C.; Curran, M. E.; Keating, M. T. A Mechanistic Link between an Inherited and an Acquird Cardiac Arrthytmia: HERG Encodes the IKr Potassium Channel. *Cell* **1995**, *81*, 299−307.
(60) Baburin, I.; Beyl, S.; Hering, S. Automated Fast Perfusion of Xenopus Oocytes for Drug Screening. *Pflugers Arch.* **2006**, *453*, 117−123.
(61) Luzhkov, V. B.; Österberg, F.; Acharya, P.; Chattopadhyaya, J.; Aqvist, J. Computational and NMR Study of Quaternary Ammonium Ion Conformations in Solution. *Phys. Chem. Chem. Phys.* **2002**, *4*, 4640−4647.
(62) Faraldo-Gómez, J. D.; Roux, B. Characterization of Conformational Equilibria through Hamiltonian and Temperature Replica-Exchange Simulations: Assessing Entropic and Environmental Effects. *J. Comput. Chem.* **2007**, *28*, 1634−1647.
(63) Choi, K.-H.; Song, C.; Shin, D.; Park, S. hERG Channel Blockade by Externally Applied Quaternary Ammonium Derivatives. *Biochim. Biophys. Acta* **2011**, *1808*, 1560−1566.
(64) Marshall, M. S.; Steele, R. P.; Thanthiriwatte, K. S.; Sherrill, C. D. Potential Energy Curves for Cation-Pi Interactions: Off-Axis Configurations Are Also Attractive. *J. Phys. Chem. A* **2009**, *113*, 13628−13632.
(65) Zhou, Y.; Morais-Cabral, J. H.; Kaufman, A.; MacKinnon, R. Chemistry of Ion Coordination and Hydration Revealed by a K+ Channel-Fab Complex at 2.0 A Resolution. *Nature* **2001**, *414*, 43−48.
(66) Perrin, M. J.; Kuchel, P. W.; Campbell, T. J.; Vandenberg, J. I. Drug Binding to the Inactivated State Is Necessary but Not Sufficient for High-Affinity Binding to Human Ether-À-Go-Go-Related Gene Channels. *Mol. Pharmacol.* **2008**, *74*, 1443−1452.
(67) Hill, A. P.; Perrin, M. J.; Heide, J.; Campbell, T. J.; Mann, S. A; Vandenberg, J. I. Kinetics of Drug Interaction with the Kv11.1 Potassium Channel. *Mol. Pharmacol.* **2014**, *85*, 769−776.
(68) Xia, M.; Shahane, S. A.; Huang, R.; Titus, S. A.; Shum, E.; Zhao, Y.; Southall, N.; Zheng, W.; Witt, K. L.; Tice, R. R.; Austin, C. P. Identification of Quaternary Ammonium Compounds as Potent Inhibitors of hERG Potassium Channels. *Toxicol. Appl. Pharmacol.* **2011**, *252*, 250−258.
(69) Knape, K.; Linder, T.; Wolschann, P.; Beyer, A.; Stary-Weinzinger, A. In Silico Analysis of Conformational Changes Induced by Mutation of Aromatic Binding Residues: Consequences for Drug Binding in the hERG K+ Channel. *PLoS One* **2011**, *6*, e28778.
(70) Karczewski, J.; Wang, J.; Kane, S. A.; Kiss, L.; Koblan, K. S.; Culberson, J. C.; Spencer, R. H. Analogs of MK-499 Are Differentially Affected by a Mutation in the S6 Domain of the hERG K+ Channel. *Biochem. Pharmacol.* **2009**, *77*, 1602−1611.
(71) Milnes, J. T.; Crociani, O.; Arcangeli, A.; Hancox, J. C.; Witchel, H. J. Blockade of HERG Potassium Currents by Fluvoxamine: Incomplete Attenuation by S6Mutations at F656 or Y652. *Br. J. Pharmacol.* **2003**, *139*, 887−898.
(72) Mitcheson, J. S. hERG Potassium Channels and the Structural Basis of Drug-Induced Arrhythmias. *Chem. Res. Toxicol.* **2008**, *21*, 1005−1010.
(73) Beyl, S.; Timin, E. N.; Hohaus, A.; Stary, A.; Kudrnac, M.; Guy, R. H.; Hering, S. Probing the Architecture of an L-Type Calcium Channel with a Charged Phenylalkylamine: Evidence for a Widely Open Pore and Drug Trapping. *J. Biol. Chem.* **2007**, *282*, 3864−3870.
(74) Imai, Y. N.; Ryu, S.; Oiki, S. Docking Model of Drug Binding to the Human Ether-à-go-go Potassium Channel Gudied by Tandem Dimer Mutant Patch-Clamp Data: A Synergic Approach. *J. Med. Chem.* **2009**, *52*, 1630−1638.
(75) Dempsey, C. E.; Wright, D.; Colenso, C. K.; Sessions, R. B.; Hancox, J. C. Assessing hERG Pore Models As Templates for Drug Docking Using Published Experimental Constraints: The Inactivated State in the Context of Drug Block. *J. Chem. Inf. Model.* **2014**, *54*, 601−612.
(76) Chen, J.; Seebohm, G.; Sanguinetti, M. C. Position of aromatic residues in the S6 dfomain, not inactivation, dictates cisapride sensitivity of HERG and eag potassium channels. *Proc. Natl. Acad. Sci. U. S. A.* **2002**, *99*, 12461−12466.

5.7 Paper #7

Molecular Determinants for Activation of Human *Ether-à-go-go-related* Gene 1 Potassium Channels by 3-Nitro-*N*-(4-phenoxyphenyl) Benzamide

Vivek Garg, Anna Stary-Weinzinger, Frank Sachse, and Michael C. Sanguinetti

Department of Physiology (V.G., M.C.S.), Nora Eccles Harrison Cardiovascular Research & Training Institute (V.G., F.S., M.C.S.), and Department of Bioengineering (F.S.), University of Utah, Salt Lake City, Utah; and Department of Pharmacology and Toxicology, University of Vienna, Vienna, Austria (A.S.-W.)

Received June 2, 2011; accepted July 8, 2011

ABSTRACT

Human *ether-à-go-go-related* gene 1 (hERG1) channels mediate repolarization of cardiac action potentials. Inherited long QT syndrome (LQTS) caused by loss-of-function mutations, or unintended blockade of hERG1 channels by many drugs, can lead to severe arrhythmia and sudden death. Drugs that activate hERG1 are a novel pharmacological approach to treat LQTS. 3-Nitro-n-(4-phenoxyphenyl) benzamide [ICA-105574 (ICA)] has been discovered to activate hERG1 by strong attenuation of pore-type inactivation. Here, we used scanning mutagenesis of hERG1 to identify the molecular determinants of ICA action. Three mutations abolished the activator effects of 30 μM ICA, including L622C in the pore helix, F557L in the S5 segment, and Y652A in the S6 segment. One mutation in S6 (A653M) switched the activity of ICA from an activator to an inhibitor, revealing its partial agonist activity. This was confirmed by showing that the noninactivating mutant hERG1 channel (G628C/S631C) was inhibited by ICA and that the addition of the F557L mutation rendered the channel drug-insensitive. Simulated molecular docking of ICA to homology models of hERG1 corroborated the scanning mutagenesis findings. Together, our findings indicate that ICA is a mixed agonist of hERG1 channels. Activation or inhibition of currents is mediated by the same or overlapping binding site located in the pore module between two adjacent subunits of the homotetrameric channel.

Introduction

The rapid delayed rectifier K⁺ current (I_{Kr}) conducted by human *ether-à-go-go-related* gene 1 (hERG1) channels is the predominant repolarizing current of cardiac action potentials in large mammals (Sanguinetti et al., 1995; Trudeau et al., 1995). Slow activation/deactivation and rapid inactivation of hERG1 channels leads to a peak in I_{Kr} during phase 3 repolarization and thus is a critical regulator of action potential duration and heart rate (Sanguinetti and Tristani-Firouzi,

This work was supported by the National Institutes of Health National Heart, Lung, and Blood Institute [Grant HL055236]; an American Heart Association (Western States Affiliate) postdoctoral fellowship; and The Austrian Science Fund [Grant P22395].

Article, publication date, and citation information can be found at http://molpharm.aspetjournals.org.
doi:10.1124/mol.111.073809.
[S] The online version of this article (available at http://molpharm.aspetjournals.org) contains supplemental material.

2006). Inherited loss-of-function mutations in hERG1 can induce torsades de pointes (TdP) ventricular arrhythmia (Curran et al., 1995) and accounts for ~40% of all cases of congenital long QT syndromes (LQTS). A gain-of-function mutation in hERG1 has been associated with the rare short QT syndrome (Brugada et al., 2004). Prolonged QT duration and TdP is most commonly an acquired disorder, often caused by block of hERG1 channels as a toxic side effect of several commonly used medications (Sanguinetti and Tristani-Firouzi, 2006). Individuals with either inherited or acquired LQTS are at an increased risk of cardiac arrhythmia and sudden death.

Congenital LQTS is commonly treated by administration of a β-adrenergic receptor blocker, and invasive and costly implantable defibrillators are used for the most severe cases. The available options for short-term drug-induced TdP are intravenous Mg²⁺, correction of any electrolyte disturbance, and discontin-

ABBREVIATIONS: hERG1, human *ether-à-go-go-related* gene 1; I_{tail}, current activated by membrane depolarization; I_{tail}, tail current; LQTS, long QT syndrome; ICA, ICA-105574, 3-nitro-*N*-(4-phenoxyphenyl) benzamide; V_t, test potential; WT, wild type; DMSO, dimethyl sulfoxide; PH/SF, pore helix/selectivity filter; TdP, torsades de pointes; P, pore; RPR260243, (3*R*,4*R*)-4-(3-(6-methoxyquinolin-4-yl)-3-oxo-propyl)-1-(3-(2,3,5-trifluorophenyl)-prop-2-ynyl)-piperidine-3-carboxylic acid; PD-118057, 2-(4-[2-(3,4-dichloro-phenyl)-ethyl]-phenylamino)-benzoic acid; MES, 4-morpholine ethanesulfonic acid; MD, molecular dynamics; NS1643, 1,3-bis-(2-hydroxy-5-trifluoromethyl-phenyl)-urea; (±)BayK 8644, 1,4-dihydro-2,6-dimethyl-5-nitro-4-[2-(trifluoromethyl)phenyl]-3-pyridinecarboxylic acid, methyl ester.

uation of the culprit drug. These options are inadequate for many patients and a mechanistic-based approach such as enhancing the cardiac repolarizing currents I_{Kr} or I_{Ks} has been proposed (Goldenberg and Moss, 2008). Several compounds that activate hERG1 channels have been fortuitously discovered during routine off-target screening for channel block (Kang et al., 2005; Zhou et al., 2005; Hansen et al., 2006; Gerlach et al., 2010). The mechanisms of action of hERG1 activators include suppression of pore (P)-type inactivation and slowed deactivation. The putative binding site for two hERG1 activators, (3R,4R)-4-(3-(6-methoxyquinolin-4-yl)-3-oxo-propyl)-1-(3-(2,3,5-trifluoro-phenyl)-prop-2-ynyl)-piperidine-3-carboxylic acid (RPR260243) (Kang et al., 2005) and 2-[4-[2-(3,4-dichlorophenyl)-ethyl]-phenylamino]-benzoic acid (PD-118057) (Zhou et al., 2005), were described recently. The binding sites are distinct and can intuitively explain the predominant mechanism of action of each specific activator. RPR260243 binds near the intracellular gate of the channel (and close to S4–S5 linker) and markedly slows deactivation while also affecting inactivation by an undefined allosteric mechanism (Perry et al., 2007). PD-118057 binds to a pocket formed by the pore helix of one subunit and nearby S6 residues of an adjacent subunit to modestly attenuate P-type inactivation and increase single channel open probability (P_o) with only minor effects on deactivation (Perry et al., 2009).

Another activator of hERG1, 3-nitro-N-(4-phenoxyphenyl) benzamide (ICA-105574) was reported to shorten action potential duration of isolated guinea pig cardiomyocytes in a concentration-dependent manner (Gerlach et al., 2010). At the maximally effective concentration, ICA induced a +180 mV shift in the voltage half-point ($V_{0.5}$) of inactivation, resulting in an increased outward current of 10 to 15 times the basal amplitude at 0 mV and slowed the rate of hERG1 current deactivation by 2-fold (Gerlach et al., 2010). To elucidate the molecular determinants for the effects of ICA, we used scanning mutagenesis of the pore region of hERG1, expression of mutant channels in *Xenopus laevis* oocytes and voltage clamp for functional analysis of mutant channels and determined the effects of ICA on inactivation-impaired hERG1 mutant channels.

Materials and Methods

Channel Mutagenesis and Expression in *X. laevis* Oocytes. HERG1 (KCNH2, isoform 1a), was cloned into the pSP64 oocyte expression vector, and mutations were introduced using the QuikChange mutagenesis kit (Agilent Technologies, Santa Clara, CA). Residues Leu553 to Ile567 in S5, Thr618 to Ser624 in PH, and Cys643 to Ile663 in S6 were mutated to alanine or cysteine (to glycine or valine for alanine residues). For some residues, alternate substitutions were introduced to enhance expression (F557L, L622C, F656T, and A661C). Amino acid substitutions at some residues expressed poorly (His562, Trp568, and Ile655) and were not studied further. cRNA was prepared by in vitro transcription with mMessage mMachine SP6 kit (Ambion, Austin, TX) after linearization of the vector plasmid with EcoRI. The isolation, culture, and injection of oocytes with cRNA were performed as described previously (Goldin, 1991; Stühmer, 1992).

Voltage Clamp. Whole-cell hERG1 currents were recorded from oocytes 1 to 4 days after cRNA injection by using the two-microelectrode voltage-clamp technique (Stühmer, 1992). Agarose-tipped microelectrodes (Schreibmayer et al., 1994) were fabricated by filling the tips of 1-mm o.d. borosilicate pipettes with 1% agarose dissolved in 3 M KCl and then back-filling with 3 M KCl. Oocytes were voltage-clamped to a holding potential of −100 mV, and 1-s pulses to a test potential (V_t) of 0 mV were applied every 15 s until current magnitude reached a steady-state level. For standard I-V relationship, step currents were elicited with 1-s pulses from −100 to +50 mV in 10-mV increments. Tail currents were measured at −70 mV. For highly inactivated mutant channels, currents were recorded from oocytes bathed in a Na$^+$-free extracellular solution with [K$^+$]$_e$ elevated to 104 mM. Step currents were elicited with 1-s pulses to a V_t that ranged from −100 to +40 mV in 10-mV increments. After each test pulse, the membrane was repolarized to −120 mV. Peak tail currents were plotted as a function of V_t measured before and after 30 μM ICA-105574 (ICA). Values were normalized to the peak values of the control currents (at +40 mV), and the data were fitted to a Boltzmann function to determine the half-point of activation ($V_{0.5}$) and the slope factor (k) of the relationship. Other voltage pulse protocols are described under *Results* and in the figure legends.

After the addition of ICA to the bathing solution, 1-s pulses to 0 mV were applied every 30 s until a new steady-state level was achieved or until 10 min. Relevant voltage protocols were then repeated in the presence of drug.

Gating currents of nonconducting G626A hERG1 mutant channels were measured using the cut-open oocyte Vaseline gap method (Bezanilla and Stefani, 1998), with pulse protocols and solutions optimized for characterizing hERG1 gating currents (Piper et al., 2003). The external solution in the top and guard chambers contained 120 mM tetramethylammonium-MES, 2 mM calcium-MES, and 10 mM HEPES, pH 7.4, with MES. The internal solution in the bottom compartment contained 120 mM potassium MES, 2 mM EDTA, and 10 mM HEPES, pH 7.4 with MES. Signals were low pass-filtered at 10 kHz and digitized at 40 kHz. Linear leak and capacitance currents were compensated by analog circuitry and subtracted online by using a p/−8 [pulse/number (P/N)] protocol.

Single hERG1 channel currents were measured in cell-attached patches as described previously (Zou et al., 1997) using standard techniques (Hamill et al., 1981) and an Axopatch 200B amplifier (Molecular Devices, Sunnyvale, CA). Electrode resistance was 8 to 12 MΩ when pipettes were filled with a solution containing 104 mM potassium gluconate, 2 mM MgCl$_2$, and 10 mM HEPES, pH 7.2 with KOH. The bath solution contained 140 mM KCl, 0.1 mM CaCl$_2$, 2 mM MgCl$_2$, and 10 mM HEPES, pH 7.2 with KOH. Single-channel current amplitudes were determined from analysis of all points amplitude histograms (pCLAMP 9; Molecular Devices) of currents filtered at 1 kHz and digitized at 5 kHz. Data are expressed as mean ± S.E. (n = number of oocytes) and analyzed by the Student's *t* test.

Molecular Modeling. The homology model of the closed-channel conformation was generated with Modeller 9v7 using the KcsA crystal structure (Protein Data Bank ID 2HVK) as a template. Modeling details, including coordinates for the open conformation, have been described previously (Stary et al., 2010).

Mutants F557L, L622C, Y652A, and A653M of hERG1 were generated in PyMOL (http://www.pymol.org/). MD simulations of closed models were performed with Gromacs version 4.5.4 (http://www.gromacs.org/) (Hess et al., 2008). Wild-type (WT) and mutant channels were embedded in an equilibrated simulation box of palmitoyloleoyl phosphatidylcholine lipids. Lipid parameters were taken from Berger et al. (1997), and the OPLS-all-atom force field (Jorgensen et al., 1996) was used for the protein. The solvent was described by the TIP4P water model (Jorgensen et al., 1983). Electrostatic interactions were calculated explicitly at a distance <1 nm, long-range electrostatic interactions were calculated at every step by particle-mesh Ewald summation (Darden et al., 1993). Lennard-Jones interactions were calculated with a cutoff of 1 nm. All bonds were constrained by using the LINCS algorithm (Hess et al., 1997), allowing for an integration time step of 2 fs. The Nose-Hoover thermostat (Nose, 1984) was used for temperature coupling (τ = 0.1 ps) and the Parrinello-Rahman barostat algorithm (Parrinello and Rahman, 1981) for pressure coupling. Conjugate gradient energy-minimization steps (1000) were performed, followed by 2 ns of restrained MD, in which the protein atoms were restrained with a force constant of 1000 kJ/mol^{-1} · nm^{-2} to their initial position, whereas ions, lipids, and solvent were allowed to move freely. Each system was then subjected to 2 × 10 ns of unrestrained MD, during which coordinates were saved every 1 ps for analysis.

Coordinates of ICA105574 were generated with GaussView 5, and the geometry was optimized with the Hartree-Fock 3–21G basis set implemented in Gaussian 09 (Gaussian Inc., Wallingford, CT) (Frisch et al., 2009). Docking was performed with the program Gold 4.0.1 (Jones et al., 1995). Coordinates of the geometric center calculated among residues Phe557, Leu622, Tyr652, and Ala653 were taken as binding site origin. The binding site radius was set equal to 10 . Operations (150,000) of the GOLD genetic algorithm were used to dock the selected compounds into the WT and mutant channels. Three snapshots (3, 6, and 8 ns) were taken from MD trajectories. The Chemscore scoring function was used to estimate free energies of binding (Gold.Chemscore.DG). Reported values are averaged over the 10 best docking poses.

Solutions and Drugs. For two-microelectrode voltage-clamp experiments, the extracellular solution contained the following: 98 mM NaCl, 2 mM KCl, 1 mM CaCl$_2$, 1 mM MgCl$_2$, and 5 mM HEPES, pH 7.6. ICA was purchased from Sigma-Aldrich (St. Louis, MO). Drug solutions were prepared fresh just before experiment by dilution of a 10 mM DMSO stock solution of ICA. Each oocyte was treated with a single concentration (30 μM) of drug unless specified otherwise.

Results

ICA Increases hERG1 Current by Suppressing Inactivation But Has No Effect on Single-Channel Conductance, Maximal Conductance, or Gating Currents. The effects of 10 and 30 μM ICA on WT hERG1 channels expressed in *X. laevis* oocytes are illustrated in Fig. 1, A and B. ICA induced a marked concentration-dependent enhancement of current magnitude. Activation of hERG1 current by 30 μM ICA reached a steady state in ~10 min at 30 μM. The fold increase in current assayed with 1-s pulses to a test potential (V_t) of +20 mV was 7.6 ± 1.1 at 10 μM and 28 ± 3.4 at 30 μM, (n = 3–8). The potency of ICA was reduced in

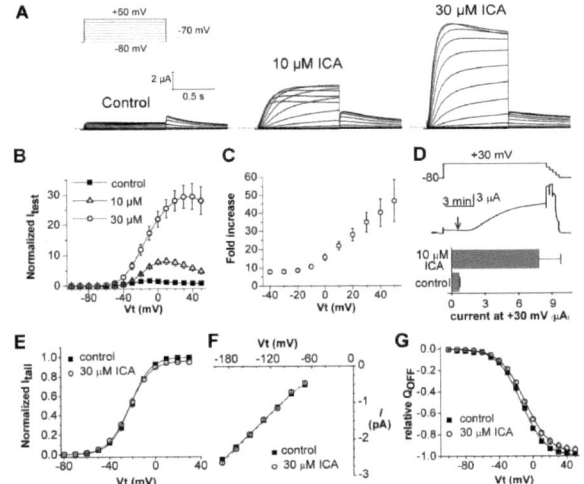

Fig. 1. Voltage-dependent activation of hERG1 current by ICA-105574 (ICA) in *X. laevis* oocytes. A, ICA increases the magnitude of hERG1 current. Currents were elicited with 1-s pulses to test potentials applied in 10-mV increments from −80 to +50 mV. Tail currents (I_{tail}) were measured at −70 mV. Voltage protocol is shown above control current traces. B, I-V_t relationships measured before (control) and in presence of 10 and 30 μM ICA (n = 3–8). Currents were normalized relative to the peak outward control current at +20 mV. C, fold increase in normalized peak outward current induced by 30 μM ICA plotted as a function of V_t (n = 8). D, ICA binds to open/inactivated state of hERG1 channels. Top, voltage-clamp protocol. Middle, corresponding current trace showing time of application of 10 μM ICA (arrow). After several minutes at +30 mV, the voltage was progressively stepped down from 0 to −80 mV in 20-mV increments. Bottom, plot of current magnitude at +30 mV before (control) and after 10 min in the presence of 10 μM ICA (n = 4). E, plot of peak I_{tail} versus V_t determined before and after 30 μM ICA in oocytes bathed in 104 mM [K$^+$]$_o$ solution. Currents were activated with 1-s pulses to a variable V_t and I_{tail} was measured at −140 mV. Peak I_{tail} magnitude was determined by fitting the current decay to a biexponential function and extrapolating the fit to the onset of membrane repolarization then normalized relative to the peak of control I_{tail} for each oocyte. Data were fitted with a Boltzmann function (smooth curves). For control, $V_{0.5}$ = −22 ± 0.5 mV, k = 8.8 ± 0.3 mV; for ICA, $V_{0.5}$ = −22.9 ± 0.3 mV, k = 9.3 ± 0.2 mV (n = 9). F, single-channel I-V_t relationship determined for cell-attached patches from oocytes bathed in control or 30 μM ICA solution. Slope conductance determined by linear fit of data for the V_t range of −100 to −180 mV was 18.1 ± 0.39 pS (n = 10) in control patches and 18.6 ± 0.28 pS (n = 10) in the presence of 30 μM ICA. G, ICA does not alter the voltage-dependence or maximal value of charge displacement ($Q_{OFF-max}$) associated with the OFF gating current of hERG1 channels. Normalized $Q_{OFF-max}$ was plotted as a function of V_t and fitted with a Boltzmann function (smooth curves). For control, $V_{0.5}$ = −14.0 ± 0.6 mV, k = 14.1 ± 0.3 mV; for ICA, $V_{0.5}$ = −9.7 ± 0.1 mV, k = 14.7 ± 0.3 mV (n = 9). The slight shift in $V_{0.5}$ was also observed for vehicle (DMSO) control (see Supplemental Fig. 1).

oocytes compared with that reported (Gerlach et al., 2010) for human embryonic kidney 293 cells (EC_{50} = 0.5 μM), probably because ICA is highly lipophilic (logP = 3.69) and accumulates in the yolk of oocytes (Witchel et al., 2002). The increase in current by ICA was associated with a marked decrease in rectification of the current-voltage (I-V_t) relationship (Fig. 1B) and an enhanced effect at more positive test potentials (Fig. 1C), effects consistent with a large drug-induced positive shift in the $V_{0.5}$ of inactivation as reported previously (Gerlach et al., 2010). ICA binds to the closed state of the hERG1 channel (Gerlach et al., 2010). In Fig. 1D, we show that ICA can also bind to channels when applied to an oocyte during a prolonged depolarizing step to +30 mV, indicating that the drug can also bind to channels in the open or inactivated state.

The voltage-dependence of the hERG1 channel conductance-voltage relationship was determined by measuring peak tail currents (I_{tail}) after a 1-s depolarizing pulse to a variable V_t. For these experiments, oocytes were bathed in 20 mM [K^+]$_o$ solution to preclude variation in I_{tail} magnitude caused by the transient local accumulation of extracellular K^+ and thus, chemical driving force associated with large outward currents and low [K^+]$_o$. When I_{tail} values were elicited at -140 mV, a potential in which recovery from inactivation is rapid and complete, ICA did not alter the voltage-dependence or G_{max} of the conductance-voltage relationship for hERG1 (Fig. 1E). This finding suggests that ICA does not cause any significant change in single-channel activity or induce recruitment of channels to the surface membrane. To confirm these expectations, we determined the effect of ICA on single-channel conductance and gating currents.

Single hERG1 channel activity was determined in cell-attached patches of oocytes using pipettes filled with 104 mM [K^+]$_o$ solution (Supplemental Fig. 1). The slope conductance for single-channel activity was not altered when 30 μM ICA was included in the pipette and bath solution (Fig. 1F). Although unlikely, ICA might also increase current magnitude by recruiting channels from a cytoplasmic store to the plasma membrane. As an indirect measure of plasma-membrane bound channel density, we determined the maximum intramembrane charge displacement ($Q_{OFF-max}$) associated with the OFF gating current. The cut-open oocyte voltage-clamp technique (Stefani and Bezanilla, 1998) was used to measure hERG1 gating currents. ICA had no effect on the kinetics of gating currents or the magnitude of the maximum intramembrane charge displacement ($Q_{OFF-max}$) associated with the OFF gating current compared with currents treated with vehicle (DMSO) (Fig. 1G and Supplemental Fig. 1). Together, these findings indicate that the ICA-induced increase in hERG1 current is not due to an increase in maximum whole-cell conductance, single-channel conductance, or an increased number of functional channels at the surface membrane and, thus, can be attributed solely to a marked attenuation of P-type inactivation (Gerlach et al., 2010).

To further explore the role of inactivation in the mechanism of action of ICA, its effects on three inactivation-impaired mutant hERG1 channels were determined (Fig. 2). S620T hERG1 channels have greatly reduced inactivation (Ficker et al., 1998) and, as expected, exhibited a greatly reduced response to ICA (only 1.3 ± 0.2-fold increase at +50 mV, n = 7). Two other point mutations (S631A, N588K) also attenuate hERG1 inactivation (Schönherr and Heinemann, 1996; Brugada et al., 2004), albeit to a lesser extent than the S620T mutation, and as expected, these channels were more sensitive to the drug compared with S620T hERG1 channels (Fig. 2). At +50 mV, 30 μM ICA increased S631A channel currents by 3.2 ± 0.2-fold (n = 4) and N588K channel currents by 2.4 ± 0.1-fold (n = 3), far less than the 47-fold enhancement observed for WT hERG1 channels (Fig. 1C). Thus, current enhancement by ICA is strongly correlated with the extent of intrinsic P-type inactivation of hERG1 channels.

Molecular Determinants for ICA Binding. Based on mutational analysis of hERG1 (Ficker et al., 1998), the entire pore module, including the pore helix/selectivity filter (PH/SF), S5 and S6 segments participate in channel inactivation.

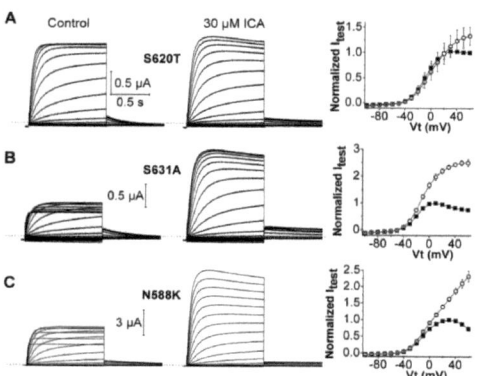

Fig. 2. Mutant channels with impaired inactivation are less sensitive to ICA. A, S620T hERG1 currents recorded before (control) and after 30 μM ICA (left and middle). Currents were elicited as described in Fig. 1A. Right, plots of mean I·V_t relationships determined before (■) and after 30 μM ICA (○). Currents (I_{test}) were measured at the end of 1-s test pulses and normalized relative to the peak outward value of control currents (n = 7). B and C, current traces (left and middle) and I-V_t relationships (right) for S631A (n = 4) and N588K (n = 3) hERG1 channels.

A

hERG1 548-AVLFL**LMCTFALIAHWLACI**WYAIGNME-575
                   ―――――――――――――――――――
                            S5

611-YVTALYF**TFSSLTS**VGFGNVSPN-633
    ――――――――――――――――
          PH

634-TNSEKIFSI**CVMLIGSLMYASIFGNVSAII**QRLY-667
                ―――――――――――――――――――――――
                              S6

B

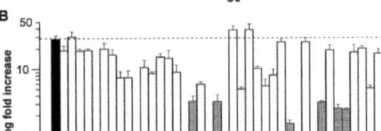

Fig. 3. Molecular determinants of hERG1 channel sensitivity to ICA. A, sequence of the S5 and pore helix-S6 domains of a hERG1 subunit. The S5, pore helix, and S6 segments are underlined, and the scanned residues are indicated by boldface text. B, bar graph summarizing fold increase in hERG1 current magnitude at +20 mV induced by 30 μM ICA for WT (■) or channels containing single point mutations as indicated. High-impact residues (≥10-fold reduction in drug effect compared with WT hERG1, $p < 0.0005$) are indicated by checkered bars. Vehicle control (DMSO) had no effect. NE, no functional expression.

To determine the molecular determinants of ICA activity, we performed scanning mutagenesis of major portions of the hERG1 pore module (Fig. 3A). A total of 44 residues were mutated, including Leu553 to Ile567 in S5, Thr618 to Ser624 in the PH/SF, and Cys643 to Ile663 in S6. The effect of ICA (30 μM for 10 min) on individual mutant channels was quantified as the fold increase in outward current measured at the end of a 1-s pulse to a V_t of +20 mV (Fig. 3B).

Nine mutations attenuated the response to 30 μM ICA by ≥10-fold and were classified as "high impact" residues (Fig. 3B). All of these mutant channels exhibited normal P-type inactivation as revealed by their typical bell-shaped I-V_t relationships. Three mutations prevented the activation of hERG1 by ICA. F557L and L622C channels were completely insensitive to 30 μM ICA (Fig. 4, A and B). ICA induced a slight but insignificant increase in outward Y652A channel currents and accelerated the rate of current deactivation; at −70 mV, the time constants for fast and slow phases of deactivation (τ_f and τ_s) were 99 ± 9.8 and 329 ± 36 ms for control, respectively, compared with 53 ± 8.3 and 221 ± 37 ms for 30 μM ICA ($n = 6$). Currents at +20 mV for five other mutant channels (F619A, S624A, F656T, N658A, and V659A) were increased <3-fold by 30 μM ICA (Supplemental Fig. 2). ICA inhibited one mutant hERG1 channel (A653M) and accelerated its rate of deactivation (Fig. 4D); at −70 mV, τ_f and τ_s were 60 ± 3.9 and 283 ± 25 ms for control, respectively, compared with 41 ± 3.7 and 216 ± 27 ms for drug ($n = 6$). In contrast to Y652A and A653M channels, ICA slowed the rate of monoexponential deactivation of WT channels at −70 mV by ~2-fold, from 413 ± 10 to 751 ± 62 ms ($n = 6$). Four mutant channels (T618A, T623A, M645C, and G648A) exhibited enhanced inactivation or very low expression and were therefore recorded in an extracellular solution containing 104 mM K^+ to accentuate the magnitude of inward I_{tail}. ICA enhanced T618A channel currents, but reduced T623A,

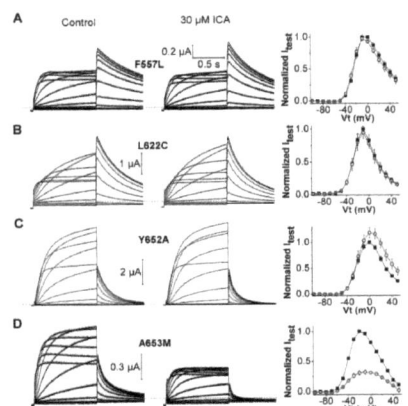

Fig. 4. Mutation of Phe557, Leu622, Tyr652, or Ala653 eliminates or reverses agonist activity of ICA on hERG1. A, left and middle, F557L hERG1 currents (elicited as described in Fig. 1A) recorded before (control) and after 30 μM ICA. Right, I-V_t relationships for currents at the end of 1-s test pulse determined before (■) and after treatment of cells with 30 μM ICA (○). Currents were normalized relative to peak outward value of control current. B to D, current traces (left and middle) and I-V_t relationships (right) for L622C (B), Y652A (C), and A653M (D) hERG1 channels.

M645C, and G648A channel currents by 30 to 50% (Supplemental Fig. 3). In summary, scanning mutagenesis identified three mutations (F557L, L622C, Y652A) that eliminated or attenuated the activator effects as well as four mutations (T623A, M645C, G648A, and A653M) that revealed an inhibitory activity of ICA.

Phe557, Leu622, and Tyr652 are likely key components of the binding site for ICA because both activator and inhibitory effects were prevented by mutation of these residues. Thr623, Ser624, Met645, Gly648, Tyr652, Phe656, and Val659 line the internal cavity, and some of these residues are important molecular determinants for pore block by hERG1 channel inhibitors (Mitcheson et al., 2000). Several of the high-impact residues identified here for ICA were found previously to be important for other hERG1 activators, including RPR260243 (Phe557, Tyr652, Asn658, and Val659) (Perry et al., 2007) and PD-118057 (Phe619, Leu622, and Met645) (Perry et al., 2009). Thus, the molecular determinants of channel activation by ICA overlap, but are not identical with other well characterized hERG1 activators and blockers.

Simulated Docking of ICA on hERG1. A homology model of the hERG1 pore module was constructed by using the KcsA (closed state) and KvAP (open state) channel structures as templates. ICA binds perpendicularly to the channel axis between two adjacent subunits of the pore module in both the open (Fig. 5A) and closed (Fig. 5C) states. A close-up view of the putative drug-binding region is depicted in Fig. 5, B and D, in which the high-impact residues identified by scanning mutagenesis and a few other residues in close con-

hERG1 Channel Activator 635

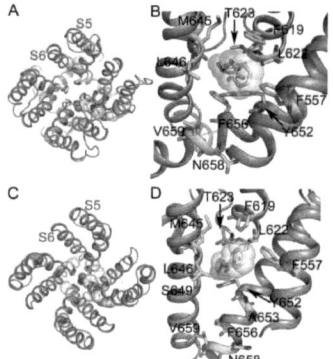

Fig. 5. ICA docked to a KvAP- (open-state) and KcsA- (closed-state) based homology model of the hERG1 pore module. A, open-state pore module of the channel (ribbons) as viewed from the extracellular space. ICA is shown in space fill. B, close-up view of the putative drug-binding region to the open-state channel. Residues designated as high-impact based on scanning mutagenesis using low $[K^+]_o$ solution (Phe557, Phe619, Leu622, Ser624, Tyr652, Ala653, Phe656, Val659, and Asn658) or high $[K^+]_o$ solution (Thr623, Met645) are labeled and are shown as stick models. Also labeled is one additional residue (Leu646) predicted to be in close contact with drug that exhibited a normal drug response when mutated to alanine. C, ICA binding to the closed-state pore module of the channel as viewed from the extracellular space. D, close-up view of the putative drug-binding region with labeling the same as in B.

tact with the drug are shown in stick mode. Except for Val659 and Asn658, all high-impact residues from the scanning mutagenesis are in close contact with ICA. In both the open and closed states of the channel, ICA resides in a hydrophobic pocket formed by Leu622, Phe619, Phe557, Tyr652, and Phe656. ICA interacts with Phe619, Phe557, and Tyr652 via π-π stacking and stabilizes the Tyr652 side chain in the down conformation (toward the cytoplasmic side of channel), in agreement with previous MD simulations (Zachariae et al., 2009). In addition, hydrogen bonds are frequently predicted between ICA and the backbone of Leu622, the side chain and/or backbone of Ser624, Thr623, and Ser649 (Supplemental Fig. 4, A and B).

The predicted binding mode of ICA is altered by the mutations that eliminated the activator effects of ICA (Supplemental Figs. 4 and 5). Atomistic molecular dynamics simulations revealed conformational changes in hERG1. For example, the F557L mutation induced side chain rotations of several residues (Leu646, Phe619, Leu622, and Thr623) that reduce the size of the lipophilic pocket. In F557L and L622C mutant channels, ICA bound to the outside of the pore module. A653M allosterically reorients the side chains of Phe619, Leu622, and Phe557, removing the lipophilic pocket. In addition, in A653M channels, hydrogen bonds to Tyr652 and Ser649 (in S6) and Ser624 and Thr623 (selectivity filter) are frequently predicted by GOLD, and in contrast to WT channels, the Tyr652 side chain frequently adopts an up-conformation. The Chemscore scoring function was used to esti-

mate free energies of binding (Gold.Chemscore.DG) of ICA and were averaged over the 10 best docking poses for WT and mutant channels. ΔG values were -40.9 and -40.54 kJ/mol for WT channels in the closed and open states, respectively, and were reduced in the closed state of the mutant channels as follows: -32.97 (F557L), -36.09 (L622C), -37.69 (Y652A), and -38.21 kJ/mol (A653M).

A Single Residue in S5 of hERG1 Determines whether ICA Is an Activator or Inhibitor. We next determined the effects of ICA on a mutant hERG1 channel that does not inactivate. Combined mutation of two residues (G628C/S631C) located near or within the selectivity filter of hERG1 completely removes channel inactivation and reduces K^+ selectivity (Smith et al., 1996). Currents conducted by G628C/S631C hERG1 channels were not augmented by ICA; instead, a 30 μM concentration of the drug decreased currents by 40% (Fig. 6, A and B). This reduction of current could result from drug binding to the central cavity receptor described for potent hERG1 blockers (Mitcheson et al., 2000). An important component of the blocker binding site is Tyr652, and mutations of this residue can greatly attenuate drug-induced block of WT channels. However, introduction of the Y652A mutation did not alter the response of G628C/S631C hERG1 channels to ICA (Fig. 6C), suggesting that current reduction is not caused by binding of ICA to the central cavity. Moreover, introduction of the F557L mutation (that prevents activator effects on WT channels) eliminated the response of G628C/S631C hERG1 channels to ICA (Fig. 6D). Together these findings suggest that ICA exerts its

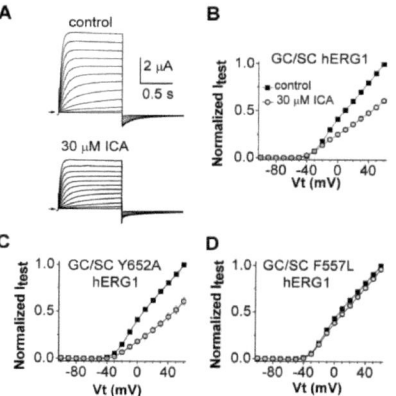

Fig. 6. Mutation of phenylalanine to leucine in S5 eliminates the inhibitory effect of ICA in a noninactivating mutant hERG1 channel. A, ICA inhibits noninactivating G628C/S631C (GC/SC) hERG1 mutant channels. Currents were elicited as described in Fig. 1A and were recorded before (control) and after 10 min of 30 μM ICA. B, I-V$_t$ relationships for currents measured at the end of 1-s test pulses before (■) and after treatment of cells with 30 μM ICA (○). Currents were normalized relative to control current at +60 mV (n = 6). C, Y652A/G628C/S631C hERG1 channels were also blocked by 30 μM ICA (○) (n = 8). D, F557L/G628C/S631C hERG1 channels are insensitive to 30 μM ICA (○) (n = 12).

Discussion

Activation of hERG1 channels by ICA is state-independent and mediated by a pronounced positive shift in the voltage dependence of P-type inactivation (Gerlach et al., 2010). P-type inactivation is caused by very subtle changes in the dynamic structure of the selectivity filter (Cuello et al., 2010b). Mutation of residues located within or near the PH/SF that impaired inactivation (S620T, S631A) caused reduced sensitivity to the drug. Inhibition of a mutant channel (G628C/S631C hERG1) with inactivation removed (Smith et al., 1996) revealed that ICA also has intrinsic antagonist activity. Phe557 (S5) and Leu622 (PH) are considered to be the most critical determinants of the binding site because single mutations of either residue prevented drug-induced changes in current magnitude and kinetics.

Voltage-gated channels can be considered analogous to intrinsically active receptors that are modulated by voltage. Viewed in this manner, ICA can behave as either an agonist or an inverse agonist of hERG1 channels. The drug is an agonist (i.e., activator) when bound to WT hERG1 channels and an inverse agonist (i.e., inhibitor) when bound to A653M or G628C/S631C hERG1 channels. Strictly speaking, the term inverse agonist would apply only if ICA bound to the same site to mediate both activator and inhibitory effects. Inhibition of noninactivating mutant hERG1 channels could simply result from pore block, mediated by binding of ICA to the central cavity as described for a plethora of hERG1 blockers (Mitcheson et al., 2000; Sanguinetti and Mitcheson, 2005) and distinct from the proposed activator site. Alternatively, both activation and inhibition of channel activity by ICA could arise from its binding to a single site as proposed for dihydropyridines that can either activate or inhibit L-type Ca^{2+} channels (Hockerman et al., 1997). For several reasons, we favor the hypothesis that ICA binds to a single site in both WT or mutant channels, and that current inhibition results from stabilization of channels in a closed (or inactivated) state rather than pore blockage per se. First, the putative activator site described for WT channels should remain intact in G628C/S631C hERG1 channels; lack of drug effect on F557L/G628C/S631C hERG1 channels supports this view. Second, if ICA bound to the well characterized central cavity site (Mitcheson et al., 2000), mutation of Tyr652 or Phe656 in hERG1 would be expected to cause an enhanced activator response rather than the observed diminished response. The side chains of these aromatic residues are the most important molecular determinants of many hERG1 blockers (Sanguinetti and Mitcheson, 2005). Mutation of Phe656 to threonine enhanced the activator effect of 1,3-bis-(2-hydroxy-5-trifluoromethyl-phenyl)-urea (NS1643) compared with WT channels (Casis et al., 2006), consistent with two binding sites for this compound: one that mediates activation, and another (in the central cavity) that mediates pore block. In contrast, F656T channels were less sensitive to ICA. Third, a conserved phenylalanine in S5 was a key determinant of both activator and inhibitory effects of ICA in hERG1 channels. Mutation of Phe557 to leucine in hERG1 (F557L) eliminated all effects of the drug. Finally, A653M hERG1 channels exhibit normal inactivation, but currents were inhibited, and deactivation was accelerated by the drug. Modeling suggests that the A653M mutation reverses the orientation of the side chain of Tyr652 (from a down to an up configuration); perhaps this interferes with the coupling between drug binding and altered inactivation gating. Whatever the underlying mechanisms, A653M in hERG1 reveals an intrinsic antagonist activity of ICA that is normally masked by its dominant activator effect. A similar switch from agonist to antagonist activity was reported for 1,4-dihydro-2,6-dimethyl-5-nitro-4-[2-(trifluoromethyl)phenyl]-3-pyridinecarboxylic acid, methyl ester [(±)BayK 8644] after mutation of two adjacent residues (Tyr1485, Met1486) in domain IV S6 of the α_{1c} L-type calcium channel (Schuster et al., 1996). It is noteworthy that the topology of the putative ICA binding site, a hydrophobic pocket located between two adjacent subunits of the pore module, is similar to the proposed dihydropyridine binding site in L-type calcium channels (Cosconati et al., 2007; Tikhonov and Zhorov, 2009) and the binding site for brevetoxins and ciguatoxin (neurotoxin receptor site 5) in voltage-gated sodium channels (Catterall et al., 2007). All of these lipophilic compounds alter channel gating and are proposed to gain direct access to the pore module via the lipid membrane.

Mutation of Tyr652 to alanine prevents the attenuation of inactivation normally caused by RPR260243 (Perry et al., 2007). However, deactivation of Y652A hERG1 channels is markedly slowed by RPR260243 (Perry et al., 2007), indicating that the mutation does not prevent drug binding. Y652A channels are also resistant to the normally pronounced effect of ICA on inactivation. However, opposite to the effects on WT channels, deactivation of Y652A channels was accelerated by ICA. The ability of both ICA and RPR260243 to affect deactivation but not inactivation of Y652A channels suggests that Tyr652 residues are required for coupling drug binding to suppression of channel inactivation. Modeling suggests that Tyr652 directly interacts with ICA but not with RPR260243 (Perry et al., 2007). Based on sequence alignments of S6 segments, Tyr652 is equivalent to Phe103 of KcsA and Ile470 of Shaker K$^+$ channels. Mutation of Phe103 (KcsA channel) or Ile470 (Shaker channel) drastically suppresses C-type inactivation, and Cuello et al. (2010a) have proposed that these key residues in S6 allosterically couple the cytosolic gate with the extracellular gate (selectivity filter), so-called bidirectional coupling. In contrast, mutations of Tyr652 in hERG1 do not appreciably alter inactivation gating (Fernandez et al., 2004). The mechanism responsible for disrupted coupling between drug binding and altered inactivation of Y652A hERG1 channels requires further study.

In summary, ICA binds to a hydrophobic pocket located between two adjacent hERG1 channel subunits, resulting in a subtle change in configuration of the selectivity filter that disrupts inactivation gating. Binding of ICA to the same or overlapping site mediates inhibition of mutant A653M and G628C/S631C hERG1 channels.

Acknowledgments

We thank Jennifer Abbruzzese for measurement of gating currents, Tobias Linder and Kirsten Knape for help with modeling figures, and Kam Hoe Ng for isolation and injection of oocytes.

Authorship Contributions

Participated in research design: Garg and Sanguinetti.
Conducted experiments: Garg and Sanguinetti.
Performed data analysis: Garg, Stary-Weinzinger, Sachse, and Sanguinetti.
Wrote or contributed to the writing of the manuscript: Garg, Stary-Weinzinger, and Sanguinetti.

References

Berger O, Edholm O, and Jähnig F (1997) Molecular dynamics simulations of a fluid bilayer of dipalmitoylphosphatidylcholine at full hydration, constant pressure, and constant temperature. *Biophys J* 72:2002–2013.
Bezanilla F and Stefani E (1998) Gating currents. *Methods Enzymol* 293:331–352.
Brugada R, Hong K, Dumaine R, Cordeiro J, Gaita F, Borggrefe M, Menendez TM, Brugada J, Pollevick GD, Wolpert C, et al. (2004) Sudden death associated with short-QT syndrome linked to mutations in HERG. *Circulation* 109:30–35.
Casis O, Olesen SP, and Sanguinetti MC (2006) Mechanism of action of a novel human ether-a-go-go-related gene channel activator. *Mol Pharmacol* 69:658–665.
Catterall WA, Cestèle S, Yarov-Yarovoy V, Yu FH, Konoki K, and Scheuer T (2007) Voltage-gated ion channels and gating modifier toxins. *Toxicon* 49:124–141.
Cosconati S, Marinelli L, Lavecchia A, and Novellino E (2007) Characterizing the 1,4-dihydropyridines binding interactions in the L-type Ca^{2+} channel: model construction and docking calculations. *J Med Chem* 50:1504–1513.
Cuello LG, Jogini V, Cortes DM, Pan AC, Gagnon DG, Dalmas O, Cordero-Morales JF, Chakrapani S, Roux B, and Perozo E (2010a) Structural basis for the coupling between activation and inactivation gates in K^+ channels. *Nature* 466:272–275.
Cuello LG, Jogini V, Cortes DM, and Perozo E (2010b) Structural mechanism of C-type inactivation in K^+ channels. *Nature* 466:203–208.
Curran ME, Splawski I, Timothy KW, Vincent GM, Green ED, and Keating MT (1995) A molecular basis for cardiac arrhythmia: HERG mutations cause long QT syndrome. *Cell* 80:795–803.
Darden T, York D, and Pedersen L (1993) Particle mesh Ewald: An N [center-dot] log(N) method for Ewald sums in large systems. *J Chem Phys* 98:10089–10092.
Fernandez D, Ghanta A, Kauffman GW, and Sanguinetti MC (2004) Physicochemical features of the HERG channel drug binding site. *J Biol Chem* 279:10120–10127.
Ficker E, Jarolimek W, Kiehn J, Baumann A, and Brown AM (1998) Molecular determinants of dofetilide block of HERG K^+ channels. *Circ Res* 82:386–395.
Frisch MJ, Trucks GW, Schlegel HB, Scuseria GE, Robb MA, Cheeseman JR and Scalmani G (2009) Gaussian 09, Revision A. 1 Gaussian, Inc., Wallingford.
Gerlach AC, Stoehr SJ, and Castle NA (2010) Pharmacological removal of human ether-a-go-go-related gene potassium channel inactivation by 3-nitro-N-(4-phenoxyphenyl) benzamide (ICA-105574). *Mol Pharmacol* 77:58–68.
Goldenberg I and Moss AJ (2008) Long QT syndrome. *J Am Coll Cardiol* 51:2291–2300.
Goldin AL (1991) Expression of ion channels by injection of mRNA into Xenopus oocytes. *Methods Cell Biol* 36:487–509.
Hamill OP, Marty A, Neher E, Sakmann B, and Sigworth FJ (1981) Improved patch-clamp techniques for high-resolution current recording from cells and cell-free membrane patches. *Pflugers Arch* 391:85–100.
Hansen RS, Diness TG, Christ T, Demnitz J, Ravens U, Olesen SP, and Grunnet M (2006) Activation of human ether-a-go-go-related gene potassium channels by the diphenylurea 1,3-bis-(2-hydroxy-5-trifluoromethyl-phenyl)-urea (NS1643). *Mol Pharmacol* 69:266–277.
Hess B, Bekker H, Berendsen HJC, and Fraaije JG (1997) LINCS: a linear constraint solver for molecular simulations. *J Comput Chem* 18:1463–1472.
Hess B, Kutzner C, van der Spoel D, and Lindahl E (2008) GROMACS 4: Algorithms for highly efficient, load-balanced, and scalable molecular simulation. *J Chem Theory Comput* 4:435–447.
Hockerman GH, Peterson BZ, Johnson BD, and Catterall WA (1997) Molecular determinants of drug binding and action on L-type calcium channels. *Annu Rev Pharmacol Toxicol* 37:361–396.
Jones G, Willett P, and Glen RC (1995) Molecular recognition of receptor sites using a genetic algorithm with a description of desolvation. *J Mol Biol* 245:43–53.
Jorgensen WL, Chandrasekhar J, Madura JD, Impey RW, and Klein ML (1983) Comparison of simple potential functions for simulating liquid water. *J Chem Phys* 79:926–935.
Jorgensen WL, Maxwell DS, and Tirado-Rives J (1996) Development and testing of the OPLS all-atom force field on conformational energetics and properties of organic liquids. *J Am Chem Soc* 118:11225–11236.
Kang J, Chen XL, Wang H, Ji J, Cheng H, Incardona J, Reynolds W, Viviani F, Tabart M, and Rampe D (2005) Discovery of a small molecule activator of the human Ether-a-go-go-Related Gene (HERG) cardiac K^+ channel. *Mol Pharmacol* 67:827–836.
Mitcheson JS, Chen J, Lin M, Culberson C, and Sanguinetti MC (2000) A structural basis for drug-induced long QT syndrome. *Proc Natl Acad Sci USA* 97:12329–12333.
Nose S (1984) A unified formulation of the constant temperature molecular dynamics methods. *J Chem Phys* 81:511–519.
Parrinello M and Rahman A (1981) Polymorphic transitions in single crystals: a new molecular dynamics method. *J Appl Phys* 52:7182–7190.
Perry M, Sachse FB, Abbruzzese J, and Sanguinetti MC (2009) PD-118057 contacts the pore helix of hERG1 channels to attenuate inactivation and enhance K^+ conductance. *Proc Natl Acad Sci USA* 106:20075–20080.
Perry M, Sachse FB, and Sanguinetti MC (2007) Structural basis of action for a human ether-a-go-go-related gene 1 potassium channel activator. *Proc Natl Acad Sci USA* 104:13827–13832.
Piper DR, Varghese A, Sanguinetti MC, and Tristani-Firouzi M (2003) Gating currents associated with intramembrane charge displacement in HERG potassium channels. *Proc Natl Acad Sci USA* 100:10534–10539.
Sanguinetti MC, Jiang C, Curran ME, and Keating MT (1995) A mechanistic link between an inherited and an acquired cardiac arrhythmia: HERG encodes the IKr potassium channel. *Cell* 81:299–307.
Sanguinetti MC and Mitcheson JS (2005) Predicting drug-hERG channel interactions that cause acquired long QT syndrome. *Trends Pharmacol Sci* 26:119–124.
Sanguinetti MC and Tristani-Firouzi M (2006) hERG potassium channels and cardiac arrhythmia. *Nature* 440:463–469.
Schönherr R and Heinemann SH (1996) Molecular determinants for activation and inactivation of HERG, a human inward rectifier potassium channel. *J Physiol* 493:635–642.
Schreibmayer W, Lester HA, and Dascal N (1994) Voltage clamping of Xenopus laevis oocytes utilizing agarose-cushion electrodes. *Pflugers Arch* 426:453–458.
Schuster A, Lacinová L, Klugbauer N, Ito H, Birnbaumer L, and Hofmann F (1996) The IVS6 segment of the L-type calcium channel is critical for the action of dihydropyridines and phenylalkylamines. *EMBO J* 15:2365–2370.
Smith PL, Baukrowitz T, and Yellen G (1996) The inward rectification mechanism of the HERG cardiac potassium channel. *Nature* 379:833–836.
Stary A, Wacker SJ, Bokharta L, Zachariae U, Karimi-Nejad Y, Aqvist J, Vriend G, and de Groot BL (2010) Toward a consensus model of the HERG potassium channel. *Chem Eur J* 6:455–467.
Stefani E and Bezanilla F (1998) Cut-open oocyte voltage-clamp technique. *Methods Enzymol* 293:300–318.
Stühmer W (1992) Electrophysiological recording from Xenopus oocytes. *Methods Enzymol* 207:319–339.
Tikhonov DB and Zhorov BS (2009) Structural model for dihydropyridine binding to L-type calcium channels. *J Biol Chem* 284:19006–19017.
Trudeau MC, Warmke JW, Ganetzky B, and Robertson GA (1995) HERG, a human inward rectifier in the voltage-gated potassium channel family. *Science* 269:92–95.
Witchel HJ, Milnes JT, Mitcheson JS, and Hancox JC (2002) Troubleshooting problems with in vitro screening of drugs for QT interval prolongation using HERG K^+ channels expressed in mammalian cell lines and Xenopus oocytes. *J Pharmacol Toxicol Methods* 48:65–80.
Zachariae U, Giordanetto F, and Leach AG (2009) Side chain flexibilities in the human ether-a-go-go related gene potassium channel (hERG) together with matched-pair binding studies suggest a new binding mode for channel blockers. *J Med Chem* 52:4266–4276.
Zhou J, Augelli-Szafran CE, Bradley JA, Chen X, Koci BJ, Volberg WA, Sun Z, and Cordes JS (2005) Novel potent human ether-a-go-go-related gene (hERG) potassium channel enhancers and their in vitro antiarrhythmic activity. *Mol Pharmacol* 68:876–884.
Zou A, Curran ME, Keating MT, and Sanguinetti MC (1997) Single HERG delayed rectifier K^+ channels in Xenopus oocytes. *Am J Physiol* 272:H1309–H1314.

Address correspondence to: Dr. Michael C. Sanguinetti, Nora Eccles Harrison Cardiovascular Research and Training Institute, Department of Physiology, University of Utah, 95 South 2000 East, Salt Lake City, UT 84112. E-mail: sanguinetti@cvrti.utah.edu

5.8 Paper #8

ICA-105574 Interacts with a Common Binding Site to Elicit Opposite Effects on Inactivation Gating of EAG and ERG Potassium Channels[S]

Vivek Garg, Anna Stary-Weinzinger, and Michael C. Sanguinetti

Nora Eccles Harrison Cardiovascular Research & Training Institute, Department of Physiology, Department of Medicine, University of Utah, Salt Lake City, Utah (V.G., M.C.S.); and Department of Pharmacology and Toxicology, University of Vienna, Vienna, Austria (A.S.-W.)

Received December 11, 2012; accepted January 14, 2013

ABSTRACT

Rapid and voltage-dependent inactivation greatly attenuates outward currents in *ether-a-go-go-related* gene (ERG) K^+ channels. In contrast, inactivation of related *ether-a-go-go* (EAG) K^+ channels is very slow and minimally reduces outward currents. ICA-105574 (ICA, or 3-nitro-N-[4-phenoxyphenyl]-benzamide) has opposite effects on inactivation of these two channel types. Although ICA greatly attenuates ERG inactivation by shifting its voltage dependence to more positive potentials, it enhances the rate and extent of EAG inactivation without altering its voltage dependence. Here, we investigate whether the inverse functional response to ICA in EAG and ERG channels is related to differences in ICA binding site or to intrinsic mechanisms of inactivation. Molecular modeling coupled with site-directed mutagenesis suggests that ICA binds in a channel-specific orientation to a hydrophobic pocket bounded by the S5/pore helix/S6 of one subunit and S6 of an adjacent subunit. ICA is a mixed agonist of mutant EAG and EAG/ERG chimera channels that inactivate by a combination of slow and fast mechanisms. With the exception of three residues, the specific amino acids that form the putative binding pocket for ICA in ERG are conserved in EAG. Mutations introduced into EAG to replicate the ICA binding site in ERG did not alter the functional response to ICA. Together these findings suggest that ICA binds to the same site in EAG and ERG channels to elicit opposite functional effects. The resultant agonist or antagonist activity is determined solely by channel-specific differences in the mechanisms of inactivation gating.

Introduction

Ether-a-go-go (EAG) K^+ channels, first described in *Drosophila* (Warmke et al., 1991), are highly expressed in the mammalian central nervous system (Ludwig et al., 1994; Martin et al., 2008) and a variety of tumors (Hemmerlein et al., 2006; Mello de Queiroz et al., 2006; Pardo et al., 1999). EAG channels activate rapidly and exhibit only a very subtle and slow form of inactivation (Garg et al., 2012). The related *ether-a-go-go-related* gene (ERG) K^+ channel was discovered by screening of a human hippocampus cDNA library (Warmke and Ganetzky, 1994), and functional analysis revealed that it activates more slowly than does EAG and undergoes a very rapid inactivation that greatly reduces channel open probability at positive potentials (Smith et al., 1996; Spector et al., 1996). Both slow (EAG) and fast (ERG) inactivation are proposed to be mediated by structural rearrangement of the selectivity filter (Stansfeld et al., 2008; Garg et al., 2012), which is commonly referred to as C- or P/C-type inactivation (Hoshi et al., 1991; Chen et al., 2000), to differentiate it from the well-characterized N-type inactivation of Kv channels (Hoshi et al., 1990).

In the human heart, ERG type 1 (hERG1, Kv11.1) channels conduct the rapid delayed rectifier K^+ current (I_{Kr}) (Sanguinetti and Jurkiewicz, 1990; Sanguinetti et al., 1995; Trudeau et al., 1995). Rapid inactivation of I_{Kr} during the plateau phase of the action potential delays repolarization and facilitates Ca^{2+} entry into the cardiomyocyte, which triggers excitation-contraction coupling. Pathologic reduction of I_{Kr}, caused either by congenital mutations in hERG1 or by block of channels as an adverse effect of many common medications is associated with a prolonged QT interval and an increased risk of cardiac arrhythmia (Sanguinetti and Tristani-Firouzi, 1996). This potentially life-threatening adverse effect prompted the now routine screening of hERG1 channel activity of compounds during the early phase of drug development programs. An unanticipated outcome of these routine screens was the discovery of compounds that activate, rather than block, hERG1 channels. Activators of hERG1 could theoretically be used to treat or prevent arrhythmia associated with congenital long QT syndrome (Zhang et al., 2012).

This work was supported by the National Institutes of Health National Heart, Lung, and Blood Institute [Grant HL055236] (to M.S.); the American Heart Association (Western States Affiliate; postdoctoral fellowship to V.G.) and The Austrian Science Fund [P22395] (to A.S.-W.).
dx.doi.org/10.1124/mol.112.084384.
[S] This article has supplemental material available at molpharm.aspetjournals.org.

ABBREVIATIONS: bEAG1, bovine *ether-a-go-go* type 1; hEAG1, human *ether-a-go-go* type 1; hERG1, human *ether-a-go-go-related* gene type 1; ICA, 3-nitro-N-(4-phenoxyphenyl)-benzamide (ICA-105574); I_{end}, current at the end of the pulse; I_{Kr}, rapid delayed rectifier K^+ current; I_{peak}, peak outward current; MD, molecular dynamics; WT, wild type.

3-Nitro-N-(4-phenoxyphenyl)-benzamide (ICA-105574, or ICA) is a recently discovered compound that shortens the duration of cardiac action potentials by inducing a dramatic shift (e.g., +183 mV at 2 uM) in the voltage dependence of hERG1 inactivation to more positive potentials (Gerlach et al., 2010). In striking contrast to its inhibition of hERG1 channel inactivation, we recently reported that ICA enhances inactivation of human EAG1 (hEAG1, Kv10.1) channels (Garg et al., 2012). In addition to differences in their response to ICA, inactivation of hERG1, but not hEAG1, is slowed by external tetraethylammonium and elevated $[K^+]_e$ (Garg et al., 2012). Despite these differences, the selectivity filter appears to serve as the inactivation gate in both hERG1 and hEAG1 channels (Smith et al., 1996; Garg et al., 2012). The disparate effects of ICA on intrinsic inactivation of such closely related Kv channels could result from ligand binding to distinct sites on the two channels or reflect inverse modulation of distinct, channel-specific modes of inactivation. Here, we use molecular modeling and analysis of mutant hEAG1 and bovine EAG1/hERG1 chimera channels to explore whether the ICA binding site in hEAG1 is homologous to the site previously reported for hERG1 (Garg et al., 2011).

Materials and Methods

Molecular Biology. Human *EAG1* (*KCNH1*; National Center for Biological Technology Information Reference Sequence: NM_002238.3) cDNA cloned into pGEMHE oocyte expression vector was kindly provided by the late Dr. Dennis Wray. *HERG1* (*KCNH2*, isoform 1a, National Center for Biological Technology Information Reference Sequence: NM_000238.2), was cloned into the pSP64 oocyte expression vector. Mutations in wild-type (WT) *hEAG1* cDNA were made using the QuikChange site-directed mutagenesis kit (Agilent Technologies, Santa Clara, CA) and were verified by DNA sequence analyses. Plasmids were linearized using NotI (pSGEMHE) or EcoR1 (pSP64). *HEAG1* cRNA was in vitro transcribed with the mMessage mMachine T7 kit (Life Technologies, Grand Island, NY). *HERG1* cRNA was prepared using the mMessage mMachine SP6 kit (Ambion, Austin, TX). cRNA was quantified using RiboGreen assay (Life Technologies).

Two-Electrode Voltage Clamp of *Xenopus* Oocytes. Procedures for harvesting oocytes from *Xenopus laevis* were as described elsewhere (Garg et al., 2012) and were approved by the University of Utah Institutional Animal Care and Use Committee. The isolation, culture, and injection of oocytes with cRNA were performed as described previously (Goldin, 1991; Stühmer, 1992). Injected oocytes were incubated for 1–5 days at 18°C in Barth's saline solution before use in voltage clamp experiments. Currents were recorded from oocytes with use of a standard two-microelectrode voltage clamp technique (Goldin, 1991; Stühmer, 1992) and agarose-cushion microelectrodes (Schreibmayer et al., 1994). A GeneClamp 500 amplifier, Digidata 1322A data acquisition system, and pCLAMP 9.0 software (Molecular Devices, Inc., Sunnyvale, CA) were used to produce command voltages and to record current and voltage signals.

Oocytes were bathed in KCM211 solution at room temperature (22–24°C). To record ionic currents, the oocyte was voltage clamped to a holding potential (V_h) of −100 mV, and 1-second pulses were applied to a test potential (V_t) of 0 mV every 10 seconds until current magnitude reached a steady-state level. During perfusion of the recording chamber with ICA solutions, the pulse interval was lengthened to 30 seconds. After currents achieved a new steady-state level in the presence of ICA, *I-V* relationships were determined if needed.

Solutions. Barth's solution contained 88 mM NaCl, 2 mM KCl, 0.41 mM CaCl$_2$, 0.33 mM Ca(NO$_3$)$_2$, 1 mM MgSO$_4$, 2.4 mM NaHCO$_3$, 10 mM HEPES, 1 mM pyruvate, and 50 mg/l gentamycin; pH was adjusted to 7.4 with NaOH. KCM211 solution contained 98 mM NaCl, 2 mM KCl, 1 mM CaCl$_2$, 1 mM MgCl$_2$, and 5 mM HEPES; pH was adjusted to 7.6 with NaOH. ICA was purchased from Sigma-Aldrich (St. Louis, MO) and AKos GmBH (Steinen, Germany) and prepared as a 10 mM stock solution in dimethyl sulfoxide. Final [ICA] was obtained by dilution of the stock solution with KCM211 immediately before use for each experiment. TEA was purchased from Sigma-Aldrich.

Data Analysis. Digitized data were analyzed off-line with pCLAMP9 (Molecular Devices), Origin 8 (OriginLab, Northhampton, MA), and Excel (Microsoft Corp., Redmond, WA) software. The concentration-effect relationship for ICA inhibition of hEAG current measured at +30 mV was fitted with a Hill equation. ICA enhanced currents at low concentrations and reduced currents at high concentrations of some mutant channels. For these mutant channels, an effective IC$_{50}$ (Table 1) was determined simply by noting the concentration that reduced control current by 50%. All data are expressed as mean ± S.E.M. (n = number of oocytes) and evaluated by an unpaired Student's t test where appropriate ($P \leq 0.05$ was considered to be a statistically significant difference).

Molecular Modeling. Homology models of the closed and open channel conformations were generated using Modeller9v9 with the KcsA crystal structure (PDB 2HVK) as a template for the closed state model and the KvAP (1ORQ) and the high resolution Mthk (PDB 3LDC) crystal structures as templates for the open conformation. Modeling details were described previously (Stary et al., 2010).

All mutant hEAG1 channels (F359L, M431F/M458L/L463M) were generated in Pymol. Molecular Dynamic (MD) simulations of open and closed models were performed using Gromacs, version 4.5.4 (Hess et al., 2008). WT and mutant channels were embedded in an equilibrated simulation box consisting of 280 dioleolylphosphatidylcholine lipids. Lipid parameters were taken from Siu et al. (Siu et al., 2008), and the TIP3P water model was used (Jorgensen et al., 1983). The amber99sb force field (Hornak et al., 2006) was used for the protein. Electrostatic interactions were calculated explicitly at a distance <1 nm and long-range electrostatic interactions were calculated at every step by particle-mesh Ewald summation (Darden et al., 1993). Lennard-Jones interactions were calculated with a cutoff of 1 nm. All bonds were

TABLE 1
Mutant hEAG1 channels with altered response to ICA

Mutation in hEAG1	Mean IC$_{50}$ for ICA ± S.E.M. μM	n	Fold increase or decrease (↓) in IC$_{50}$
(WT)	0.44 ± 0.03	5	—
V356A**	0.08 ± 0.009	3	5.5 ↓
V356E**	28.8 ± 2.6	3	65
F359A*	29.6 ± 9.0	3	67
F359L	>30	4	>68
L427A**	0.17 ± 0.01	3	2.6 ↓
M431A*	1.9 ± 0.34	5	4.3
L434A**	8.5 ± 0.08	5	19
*L434C***	4.8 ± 0.2	3	11
M457A	>30	4	>68
M458A**	20.2 ± 1.2[b]	4	46
M458E[a]	0.45 ± 0.08	3	no change
L462A**	0.05 ± 0.005	3	8.6 ↓
L463A**	41.2 ± 8.2[b]	3	94
Y464A**	0.07 ± 0.014	3	6.3 ↓
I467A**	1.1 ± 0.05	3	2.5
*I467E***	24.8 ± 4.1[b]	3	56
F468A**	3.1 ± 0.48	4	7

hEAG1, human *ether-a-go-go* type 1; ICA, -nitro-N-(4-phenoxyphenyl)-benzamide (ICA-105574); WT, wild-type.
[a] NS, nonsignificant ($P > 0.05$).
[b] Effective IC$_{50}$.
* $P \leq 0.01$.
** $P < 0.001$ compared with WT.

constrained by using the LINCS (Linear Constraint Solver) algorithm (Hess et al., 1997), allowing for an integration time step of 2 femtoseconds. The Nose-Hoover thermostat (Nose, 1984) was used for temperature coupling (τ = 0.1 picoseconds), and the Parrinello-Rahman barostat algorithm (Parrinello and Rahman, 1981) was used for pressure coupling. One thousand conjugate gradient energy-minimization steps were performed, followed by 2 nanoseconds of restrained MD, in which the protein atoms were restrained with a force constant of 1000kJ × mol^{-1}nm^{-2} to their initial position, and ions, lipids, and solvent were allowed to move freely. Each system was then subjected to 50 nanoseconds of unrestrained MD, during which coordinates were saved every 10 picoseconds for analysis.

Coordinates of ICA were generated using Gaussview 5, and the geometry was optimized with the Hartee-Fock 3-21G basis set implemented in Gaussian09 (Frisch et al., 2009). Docking was performed using the program Gold 4.0.1 (Cambridge DataCentre, Cambridge, UK) (Jones et al., 1995). Coordinates of the geometric center calculated among residues F359, L434, M431, Y464, and F468 were taken as binding site origin. The binding site radius was set equal to 10 Å; 150,000 operations of the GOLD genetic algorithm were used to dock the selected compounds into the WT and mutant channels. Three snapshots (15, 33, and 50 nanoseconds) were taken from MD trajectories. The stability of the predicted binding modes of ICA to WT hERG1 channels in open and closed conformation was confirmed in 50 nanosecond MD simulations as described previously (Knape et al., 2011).

Results

ICA Binds to the Pore Domain of hERG1 and hEAG1 Channels to Exert Opposite Effects on Inactivation.

The effects of ICA on WT hERG1 and hEAG1 channels expressed in *Xenopus* oocytes are shown in Fig. 1. Channels were activated with a 10-second pulse to +30 mV from a holding potential of −100 mV. As reported previously (Gerlach et al., 2010; Garg et al., 2011, 2012), ICA caused a marked and concentration-dependent enhancement of hERG1 current (Fig. 1A), but inhibited hEAG1 current (Fig. 1B) by reducing both the initial peak outward current (I_{peak}) and inducing a time-dependent decay of current during the pulse. Inhibition of hEAG1 current by ICA is caused by an enhancement of intrinsic inactivation and is not attributable to open channel block (Garg et al., 2012). The inhibitory effect of ICA on hEAG1 was more potent than was the activation effect on hERG1 channels. The IC$_{50}$ for ICA inhibition of hEAG1 was 1.38 ± 0.04 µM for I_{peak} and 0.44 ± 0.03 µM for current at the end of the pulse, I_{end} (n = 5) (Fig. 1C).

Using scanning mutagenesis and functional analysis of mutant hERG1 channels, we recently proposed that ICA binds to a hydrophobic pocket formed by the S5/pore helix/S6 of one subunit and S6 segment of an adjacent channel subunit

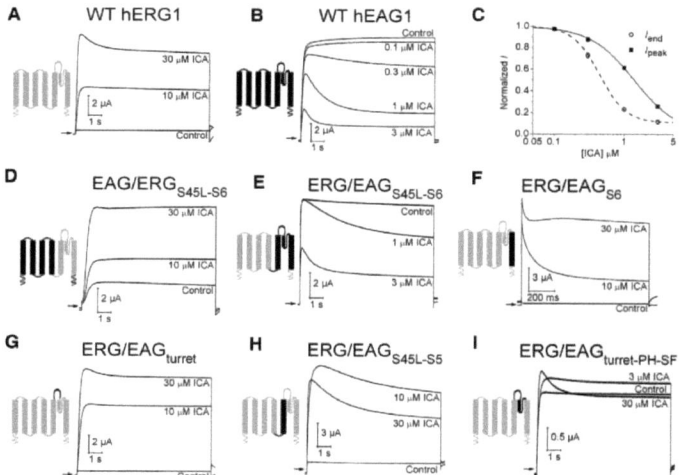

Fig. 1. The pore domain of EAG1 and ERG1 channels determines functional response to ICA. (A) ICA activates WT hERG1 channel current. (B) ICA inhibits WT hEAG1 channel current. For traces shown in A and B, currents were activated with 10-second pulses to +30 mV from a holding potential of −100 mV. (C) [ICA]-response relationship for inhibition of hEAG1 channel currents (n = 5). ICA inhibited I_{end}, current at the end of a 10-second pulse to +30 mV (IC$_{50}$ = 0.44 ± 0.03 µM; n$_H$ = 2.3) more potently than I_{peak}, peak outward current (IC$_{50}$ = 1.38 ± 0.04 µM; n$_H$ = 1.4). (D–I) Effect of ICA on EAG1/ERG1 chimera channels. In each panel, the diagram indicates regions of subunits contributed by bEAG1 (black) and hERG1 (gray). Currents were elicited with 10-second pulses to +30 mV from a holding potential of −100 mV. The fold-change in I_{end} induced by ICA for each chimera channel was as follows: (D) EAG/ERG$_{S45L-S6}$ (1.9 ± 0.03 at 10 µM; 3.6 ± 0.6 at 30 µM, n = 3). (E) ERG/EAG$_{S45L-S6}$ (0.58 ± 0.07 at 1 µM; 0.24 ± 0.04 at 3 µM, n = 4). (F) ERG/EAG$_{S6}$ (9.6 ± 3.5 at 10 µM; 41 ± 11 at 30 µM, n = 4). (G) ERG/EAG$_{turret}$ (15 ± 1.9 at 10 µM; 26 ± 0.3 at 30 µM, n = 3). (H) ERG/EAG$_{S45L-S5}$ (9.7 ± 1.6 at 10 µM; 6.1 ± 1.0 at 30 µM, n = 3). (I) ERG/EAG$_{turret-pore helix-selectivity filter}$ (% change in I_{end}: 9 ± 3 at 3 µM; 4 ± 2 at 10 µM; -3 ± 3 at 30 µM, n = 3).

(Garg et al., 2011). To identify the region of the EAG1 channel responsible for the inhibitory effect of ICA, we first used an unbiased approach and examined several previously characterized chimera channels (Ficker et al., 1998) constructed from bovine EAG1 (bEAG1) and hERG1. The amino acid sequences of bovine and human EAG1 are 97% identical for the entire proteins and 100% identical in the S1–S6 regions. First, the role of the pore module was investigated. ICA was an agonist when the pore region (S45 linker-S6) of the chimera channel was contributed by hERG1 (Fig. 1D), but was an antagonist when the same pore region was supplied by bEAG1 (Fig. 1E). Next, ICA was tested on chimeras in which only a limited region of the pore domain of hERG1 was replaced by their bEAG1 counterpart. When only the S6 segment (Fig. 1F) or the turret (Fig. 1G) was supplied by bEAG1, ICA acted as an agonist. Finally, when the S45 linker-S5 (Fig. 1H) or the turret-pore helix-selectivity filter (Fig. 1I) was supplied by bEAG1, the effects of ICA were biphasic; 30 μM ICA induced or enhanced the rate of slow inactivation and caused a smaller increase in current, compared with 10 μM ICA. Together, the findings obtained with hERG1/bEAG1 chimeras indicate that the pore domain (S5–S6) region determines the channel-specific response to ICA.

ICA Is a Mixed Agonist of a Fast-Inactivating Mutant hEAG1 Channel. Two Ser residues, located on either side of the selectivity filter, are key determinants of fast P/C-type inactivation in hERG1 (Suessbrich et al., 1997). In bEAG1 channels, combined mutation of the residues located in the homologous position as these two Serines induces fast inactivation (Ficker et al., 1998). Introduction of the same two mutations (T432S/A443S) into hEAG1 also induces a rapid, time-dependent decay of outward current (Fig. 2A), and similar to P/C-type inactivation of hERG1 (Smith et al., 1996), its rate is slowed by extracellular tetraethylammonium (10 mM) or elevated $[K^+]_e$ (Supplemental Fig. 1). In contrast, the rate of the comparatively very slow inactivation of hEAG1 is unaffected by tetraethylammonium or changes in $[K^+]_e$ (Garg et al., 2012). When activated by a 0.4-second pulse to +30 mV, 3 μM ICA reduced I_{peak} and the inactivating component of outward T432S/A443S hEAG1 current, but doubled the magnitude of I_{end} (Fig. 2A). At 30 μM, ICA eliminated fast inactivation (Fig. 2A). ICA (1–30 μM) reduced I_{peak} in a concentration-dependent manner (Fig. 2B), but exhibited a biphasic (mixed agonist) effect on I_{end} and reduced rectification of the I_{end}-V relationship (Fig. 2C). At 30 μM, ICA inhibited I_{peak} (all potentials) and I_{end} (at test potentials < +10 mV). The [ICA]-response relationship for I_{peak} and I_{end} at a single voltage (+30 mV) for T432S/A443S hEAG1 channels is compared with WT channels in Fig. 2D. Together, these findings suggest that the multiple effects of ICA on T432S/A443S hEAG1 channels represent the sum of two opposing actions: reduced fast (ERG1-like) inactivation to increase current plus inhibition, perhaps mediated by open channel block to reduce current. We next explore whether the opposite effects of ICA on fast versus slow modes of inactivation gating are mediated by ICA binding to the same or distinct sites in hERG1 and hEAG1 channels.

Molecular Determinants of ICA Interaction with hEAG1 Channels. Simulated docking of ICA to molecular models of the hEAG1 pore module was performed, and the findings were compared with those from our previous model of ICA bound to the hERG1 channel. Homology models of hEAG1 were constructed using the KvAP and MthK crystal structures (Jiang et al., 2003; Ye et al., 2010) as template for the open state and the KcsA crystal structure (Doyle et al., 1998) as template for the closed state. In hERG1, ICA was predicted to be oriented perpendicular to the axis of the S5 and S6 segments in both the closed and the open state dockings, with the nitro group facing toward the pore (Garg et al., 2011). For hEAG1, ICA also favored a perpendicular orientation in the open state, but the nitro group faces away from the pore and toward the lipids that surround the pore

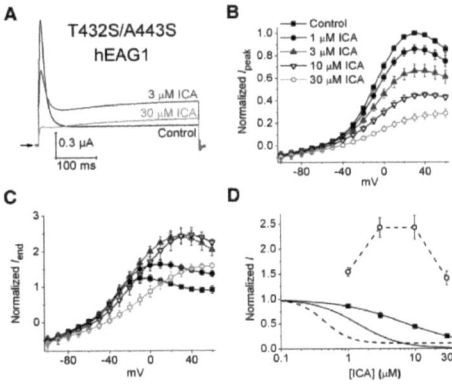

Fig. 2. ICA has dual effects on fast-inactivating T432S/A443S hEAG1 channels. (A) Representative traces showing the effect of 3 and 30 μM ICA on current measured at a test potential of +30 mV. Currents were elicited from a holding potential of −60 mV. (B and C) Normalized I_{peak}-V relationships (B) and I_{end}-V (C) relationships measured before (control) and in the presence of indicated [ICA] (n = 5). (D) [ICA]-response relationships for I_{peak} (solid curves) and I_{end} (dashed curves) measured at a single test potential (+30 mV) for T432S/A443S (data points, ■ and ○) and WT hEAG1 (curves without data points). The IC$_{50}$ for I_{peak} was 7.6 ± 0.9 μM for T432S/A443S hEAG1 (n = 5).

module (Fig. 3, A–C). In the closed state of hEAG1, ICA is orientated parallel to S5 and S6 (Fig. 3, D–F). Simulated docking of ICA to the open states of hERG1 and hEAG1 are compared in Supplemental Fig. 2. In hEAG1, ICA resides in a hydrophobic pocket formed by residues Met431, Leu434, Met458, Tyr464, Ile467, and Phe359. The location of this binding pocket is quite similar to that previously described for hERG1 (Garg et al., 2011). However, sequence differences between hERG1 and hEAG1 lead to different-shaped binding pockets for ICA (Supplemental Fig. 2, C and E). ICA protrudes deeply into the cleft between two adjacent subunits in the hERG1 channel (Supplemental Fig. 2A). By contrast, in hEAG1, Tyr464 forms a barrier at the S6-S6 interface, leading to a shallower binding mode for ICA (Fig. 3; Supplemental Fig. 2D). ICA does not form π-π stacking interactions with Tyr464 in either the closed or the open state of hEAG1 and, thus, might not be able to stabilize the phenyl group in the down conformation, as previously suggested for hERG1 (Garg et al., 2011). In addition, stabilizing hydrogen bonds predicted between ICA and selectivity filter residues in hERG1 are lacking in hEAG1.

Considering both open and closed hEAG1 model simulations, 11 residues are predicted to be in close proximity to ICA: Leu427, Met431, and Leu434 in the pore helix; Val356 and Phe359 in S5; Leu463, Tyr464, Ile467, and Phe468 in S6 of one subunit; and Met457 and Met458 in the S6 of an adjacent subunit. To corroborate the modeling results, we mutated to Ala each of these 11 residues. Seven of the 11 Ala substitutions reduced the sensitivity (increased IC_{50}) of the mutant channel to ICA by >4-fold, compared with WT hEAG1 (Table 1; Supplemental Fig. 3). The I467A mutation increased the IC_{50} by 2.5-fold, whereas V356A, L427A, and Y464A mutations reduced IC_{50} by 2.5–6-fold. Val356, Leu427, and Ile467 are predicted to interact with ICA in the closed but not the open state. Because of the poor expression of L427A mutant channel (requiring 100 times more cRNA and longer incubation time in comparison with WT hEAG) and its location near the selectivity filter (P/C-type inactivation gate), Leu427 residue was not analyzed further. To explore the potential significance of Ile467 and Val356 to ICA binding, each was mutated to the more perturbing Glu. Accordingly, both V356E and I467E mutations increased the IC_{50} for ICA by >55-fold (Supplemental Fig. 3; Table 1). Thus, mutagenesis confirmed the importance of all the residues predicted by molecular modeling to interact with ICA in either the open or the closed state of the hEAG1 channel.

Some of the molecular determinants of ICA binding to hEAG1 align with those previously described for hERG1. For example, mutation of Phe557, Phe619, Leu622, and Phe656 in hERG1 reduced the sensitivity to the activator effect of ICA (Garg et al., 2011); similarly, mutation of the corresponding residues to Ala in hEAG1 (Phe359, Met431, Leu434, and Phe468) reduced the inhibitory action of ICA. However, many homologous mutations yielded incongruous findings for the two channels. First and of most importance, the Y652A hERG1 channel is nearly insensitive to activation by ICA (Garg et al., 2011), whereas the homologous Y464A hEAG1 channel is more sensitive to inhibition (more inactivated) by ICA (Garg et al., 2012). Second, although the mutations L463A and M458A decreased the sensitivity of hEAG1 channels (IC_{50} increased by 94-fold and 46-fold, respectively; Table 1) and V356A enhanced sensitivity to ICA by 6-fold, the corresponding mutations in hERG1 (M651A, L646A, and M554A) were previously reported to not alter the response to ICA (Garg et al., 2011). Third, F557L hERG1 channels are insensitive to activation by ICA, whereas the corresponding F359L hEAG1 channel is activated by ICA (Garg et al., 2012). We investigated these notable differences by further examining three of these key residues in hEAG1: Met458 and Tyr464 in S6 and Phe359 in S5.

Simulated dockings predicted that both Met458 in hEAG1 and the homologous residue Leu646 in hERG1 are in close

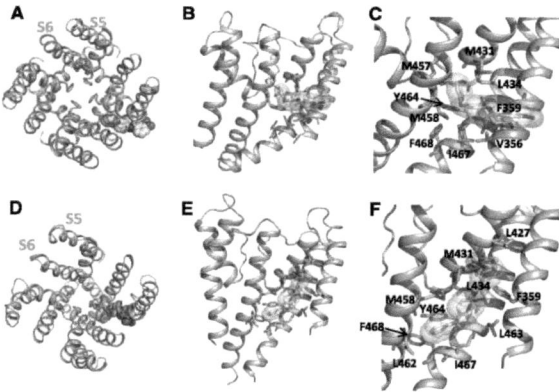

Fig. 3. Models of ICA docked to the open and closed state of the hEAG1 pore module. (A) Single ICA molecule (shown in space-fill) docked to the open state model of the complete pore module as viewed from the extracellular space. (B) Side view of the S5-S6 regions of two adjacent hEAG1 subunits of an open state channel. ICA molecule is shown in transparent space-fill; interacting residues are shown as green sticks. (C) Close-up view of panel B with interacting residues labeled. (D–F) ICA bound to the closed state model of the hEAG1 pore module.

proximity to ICA. Nonetheless, substitution of Leu646 to either Ala (Garg et al., 2011) or Glu (Supplemental Fig. 4) did not alter hERG1 channel sensitivity to ICA. As noted above, M458A mutation in hEAG1 reduced ICA sensitivity; however, similar to L646E in hERG1, M458E did not alter hEAG1 sensitivity to ICA (Supplemental Fig. 4; Table 1), presumably because the acidic side chain of Glu is repelled from the hydrophobic pocket and, thus, does not affect ICA binding. In hERG1, three specific point mutations, Y652A (S6), F557L (S5), and L434C (pore helix), rendered the channel insensitive to 30 μM ICA, consistent with molecular modeling predictions (Garg et al., 2011). The effect of multiple ICA concentrations on mutant channels harboring the homologous substitutions in hEAG1 (Y464A, F359L, L434C) were examined (Fig. 4, A–D). Consistent with Leu434 contributing to ICA binding, L434C hEAG1 was less inhibited by ICA (IC$_{50}$ = 4.8 ± 0.2 μM), compared with WT channels (Fig. 4, A and D). As we reported previously, Y464A promotes and ICA accentuates prominent inactivation from an open state (Fig. 4, B and D), whereas F359L (Fig. 4C) appears to promote and ICA reverses inactivation from closed states (Garg et al., 2012). Molecular modeling predicts that the F557L mutation in hERG1 excludes ICA from interaction with its hydrophobic binding pocket (Garg et al., 2011), consistent with its insensitivity to the drug. In contrast, molecular modeling predicts that the hydrophobic pocket in F359L hEAG1 channels can accommodate ICA (Supplemental Fig. 5), albeit in a different orientation, compared with the WT hEAG1 channel. Together, molecular modeling predictions and functional analysis of many mutant channels indicate that ICA modulates inactivation gating of both hEAG1 and hERG1 channels by interacting with the same hydrophobic pocket defined by the S5-pore helix-S6 region of one subunit and S6 of an adjacent subunit. We next sought to determine whether the functional effect of ICA could be reversed (i.e., switched from inhibitor to agonist) if the putative hydrophobic pocket of hEAG1 was modified by mutagenesis to mimic the pocket present in hERG1 channels.

Introducing Putative hERG1 Binding Pockets into hEAG1 Does Not Alter Response to ICA. The S5-pore helix-S6 regions of hERG1 and hEAG1 are composed of 79 residues, and protein sequence alignment (Fig. 5A) indicates several differences between the two channels, including 8 residues in S5, 5 residues in the pore helix, and 14 residues in S6. However, in the putative ICA binding pocket defined by docking simulations using the open state models of hERG1 (Garg et al., 2011) and hEAG1 (Fig. 3), only one residue in the pore helix and two residues in S6 differ between the two channels (Fig. 5A). Three amino acid substitutions were introduced into hEAG1 (M431F in the pore helix; M458L and L463M in S6) to match the corresponding residues in hERG1. The resulting triple-mutant (M431F/M458L/L463M) channel retained WT hEAG1 biophysical properties and response to ICA (Fig. 5A), including a similar IC$_{50}$ value for inhibition of I_{end} (0.49 ± 0.15 μM, n = 3; Fig. 5C). Can the putative ICA binding site in hEAG1 be adequately recapitulated in hEAG1 by just three amino acid substitutions? Molecular modeling suggests remarkable similar binding modes of the triple hEAG1 channel, compared with the WT hERG1 channel. Simulated docking predicts that ICA binds perpendicular to the axis of the S5 and S6 segments in both the closed and the open state, and the nitro group faces toward the pore (Fig. 6), similar to the orientation of ICA in hERG1 (Garg et al., 2011). Furthermore, mutations M431F/M458L/L463M render the shape of the binding site to be more hERG1-like, allowing the drug to protrude deeply into the cleft formed by the interface of two adjacent subunits. Together, these modeling and experimental findings support the notion that intrinsic differences in the mechanisms of slow versus fast inactivation gating, and not differences in the binding site, determines whether ICA is a channel antagonist (hEAG1) or agonist (hERG1).

Discussion

ICA Binds to a Common Site of EAG1 and ERG1 Channels to Exert Opposite Effects on Inactivation. ICA inhibits outward K^+ hEAG1 channel currents by enhancing slow inactivation (i.e., it is an agonist of intrinsic slow inactivation gating). In contrast, ICA enhances outward hERG1 K^+ channel currents by inhibiting inactivation (i.e., it is an antagonist of intrinsic fast inactivation gating). Despite the opposite functional response to ICA, analysis of chimera ERG/EAG channels and multiple mutant channels clearly establish that the compound binds to a similar region, in a hydrophobic cleft between two adjacent subunits of the pore module in both hERG1 and hEAG1. A recent MD simulation study of hERG1 proposed that the binding pocket for ICA is located between the pore helices of two adjacent subunits and that the selectivity filter adopts a collapsed conformation in the inactivated state, precluding entry of the compound into the pocket (Kopfer et al., 2012). However, this binding mode does not include interaction with residues in S5, including

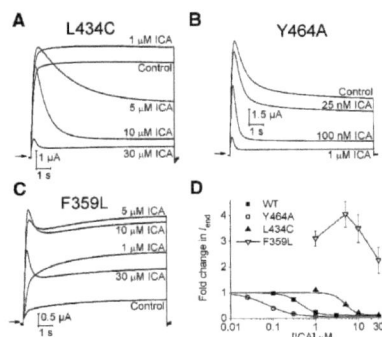

Fig. 4. Concentration-dependent effects of ICA on L434C, Y464A, and F359L hEAG1 channels. (A) Effect of ICA (1–30 μM) on L434C hEAG1 channel currents measured in response to 10-second depolarization to +30 mV. (B) Effect of ICA (25 nM to 1 μM) on Y464A hEAG1 channel currents during 10-second pulse to +30 mV. (C) Biphasic response of F359L hEAG1 channels to ICA (1–30 μM) during 10-second pulse to +30 mV. (D) [ICA]-response (normalized I_{end}) relationships for indicated WT and mutant hEAG1 channels. The IC$_{50}$ for I_{end} was 0.07 ± 0.01 μM for Y464A (n = 3), 4.8 ± 0.2 μM for L434C (n = 3) and >30 μM for F359L (n = 4) hEAG1 channels. WT data are same as that plotted in Fig. 1C.

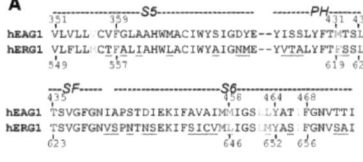

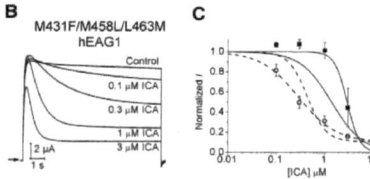

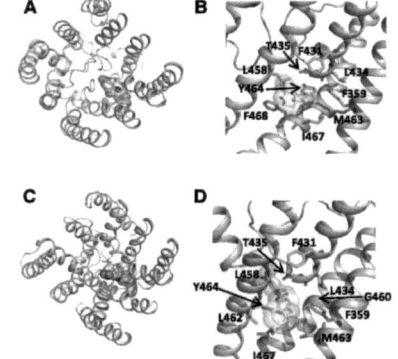

Fig. 5. ICA-induces inactivation of mutant hEAG1 channel with putative drug-binding pocket engineered to mimic hERG1 channel. (A) Amino acid sequence alignment of S5 and pore helix (PH)/selectivity filter (SF)/S6 region of hEAG1 and hERG1. Nonconserved amino acids are underlined. Conserved residues (blue) and nonconserved residues (red) predicted to line the hydrophobic ICA binding pocket in hERG1 are highlighted. Residues colored green were predicted to contribute to ICA binding site in closed state of hEAG1 (but not predicted to interact with hERG1 in closed or open state). (B) Concentration-dependent inhibition of M431F/M458L/L463M hEAG1 channel currents by ICA. Currents were elicited with 10-second pulses to +30 mV. (C) [ICA]-response relationships for I_{peak} (solid curve) and I_{end} (dashed curve) quantified as fold change in current measured at a test potential of +30 mV for M431F/M458L/L463M (data points, ■ and ○) and WT hEAG1 (curves only, replotted from Fig. 1C). The IC_{50} was 0.49 ± 0.15 μM for I_{end} and 2.96 ± 0.82 μM for I_{peak} (n = 4).

Fig. 6. Molecular models of ICA docked to M431F/M458L/L463M hEAG1 channel pore module. (A and B) Open state model. (C and D) Closed state model.

F557, a residue that we find to be of particular importance in modification of channel gating by ICA. Because both channels are homotetramers, there could be four identical ICA binding sites on each channel. Consistent with multiple binding sites, activation of hERG1 channels by ICA exhibits strong co-operativity, with a Hill coefficient of 3.3 estimated for the concentration-response relationship (Gerlach et al., 2010).

Slow inactivation of hEAG1 channels is modulated by a proposed interaction among three residues in close proximity and located in the S5 (Phe359), pore helix (Leu434) and S6 (Tyr464) of each subunit (Garg et al., 2012). In WT channels, inactivation is very slow and barely detectable but is greatly enhanced by ICA or mutations of Tyr464. Y464A hEAG1 channels exhibit far greater intrinsic slow inactivation than do WT channels, and ICA accentuates this altered mode of gating. Inactivation of Y464A channels can be prevented (WT gating restored) by introducing a second mutation of either Leu434 or Phe359 (Garg et al., 2012). In contrast to Tyr464, multiple mutations of Leu434 or Phe359 do not alter the biophysical properties of hEAG1 (Garg et al., 2012), but do affect the response to ICA. F359A and L434A/C reduce the efficacy of ICA to induce inactivation, whereas highly in-activated F359L channels are activated by ICA. Together, these findings suggest that ICA directly affects the molecular machinery of slow inactivation in hEAG1 channels.

F359A and F359L hEAG1 channels exhibited altered responses to ICA. F359A channels were less sensitive to inhibition by ICA (67-fold increase in IC_{50}), indicating a reduced binding affinity. In addition, ICA reduced F359A channel currents without inducing the prominent time-dependent decay of current during a depolarizing pulse seen with WT channels. We interpret this later effect to indicate that ICA enhances closed (but not open) state inactivation of F359A channels. Inhibition of F359A channels could also result from open channel block; however, we have previously presented extensive evidence that ICA induces both closed and open state inactivation of WT channels with no evidence of open channel block (Garg et al., 2012). On the basis of our molecular modeling results, ICA binds similarly to the open state of WT and F359L hEAG1 channels (Supplemental Fig. 2, D and E, WT; Supplemental Fig. 5, A and B, F359L). However, unlike WT channels, the activation of F359L channels was biphasic: currents were activated at all concentrations examined (1–30 μM) and peaked at 3 μM. ICA concentrations >3 μM led to progressively less activation that was accompanied by progressively more extensive time-dependent decay of outward currents (indicative of enhanced open channel inactivation). On the basis of injection of oocytes with equivalent amounts of cRNA, F359L hEAG1 channel currents are much smaller than WT channel currents, suggesting that these mutant channels are either highly inactivated or have a lower than normal single channel open probability. As discussed previously (Garg et al., 2012), we propose that ICA-mediated increase in F359L channel currents may be caused by a reduced rate of closed to inactivated state transitions.

Modification of channel gating has also been proposed as the mechanism responsible for activation of KCNQ2-5 (Kv7.2-Kv7.5) channels by retigabine, an anticonvulsant drug that shifts the voltage dependence of activation to more negative potentials. The putative binding site for retigabine is a hydrophobic binding pocket (Schenzer et al., 2005; Wuttke et al., 2005; Lange et al., 2009) located in the same region described here for ICA binding to ERG and EAG channels. The primary molecular determinants of retigabine binding in KCNQ3 are Trp265 (S5), Leu314 (pore helix), and Leu338 (S6), homologous to key components of the ICA binding site in hEAG1 (Phe359, Leu434, Met458) and hERG1 (Phe557, Leu622, Leu646). Moreover, mutation of the aromatic residue in S5 to Leu renders hERG (F557L) channels as insensitive to ICA and KCNQ3 (W265L) channels as insensitive to retigabine. The homologous mutation in hEAG1 (F359L) reversed the effect of ICA from antagonist to agonist activity, and the reverse mutation at the corresponding residue in KCNQ1 (L266W) leads to inhibition in a channel that is normally insensitive to retigabine (Schenzer et al., 2005). Another interesting analogy between EAG and KCNQ (specifically, KCNQ1) channels is that, in both, a tripartite mode of inactivation gating has been proposed, involving specific residues in the S5, pore helix, and S6 (Seebohm et al., 2005; Garg et al., 2012). Thus, gating of the selectivity filter in multiple, unrelated Kv channels is modulated by binding of lipophilic compounds to the hydrophobic cleft situated between two adjacent subunits in the pore module.

Clinical Relevance. Treatments for congenital and acquired long QT syndrome are limited. The recent discovery of several compounds that activate hERG1 channels initiates a promising pathway toward development of genotype-specific therapy for this life-threatening disorder. As a consequence of its profound inhibition of inactivation, ICA increases the magnitude of outward hERG1 currents more than has been observed for other activators, such as RPR260243 (Kang et al., 2005), PD-118057 (Perry et al., 2007), or NS1643 (Grunnet et al., 2011; Hansen et al., 2006). A hERG1 activator that has more modest effect on current magnitude than ICA would be less prone to induce excessive shortening of action potentials and avoid the potential conversion of long to short QT syndrome. In addition, our findings warn that hERG1 agonists may also affect the gating of highly related hEAG1 channels with potential functional consequences in the central nervous system.

Acknowledgments

The authors thank Alison Gardner and Jennifer Abbruzzese for technical assistance. The computational results were achieved using the Vienna Scientific Cluster.

Authorship Contributions

Participated in research design: Garg, Stary-Weinzinger, Sanguinetti.
Conducted experiments: Garg.
Performed data analysis: Garg, Stary-Weinzinger, Sanguinetti.
Wrote or contributed to the writing of the manuscript: Garg, Stary-Weinzinger, Sanguinetti.

References

Chen J, Avdonin V, Ciorba MA, Heinemann SH, and Hoshi T (2000) Acceleration of P/C-type inactivation in voltage-gated K(+) channels by methionine oxidation. *Biophys J* 78:174–187.
Darden T, York D, and Pedersen L (1993) Particle mesh Ewald: An $N \cdot \log(N)$ method for Ewald sums in large systems. *J Chem Phys* 98:10089–10092.
Doyle DA, Morais Cabral J, Pfuetzner RA, Kuo A, Gulbis JM, Cohen SL, Chait BT, and MacKinnon R (1998) The structure of the potassium channel: molecular basis of K$^+$ conduction and selectivity. *Science* 280:69–77.
Ficker E, Jarolimek W, Kiehn J, Baumann A, and Brown AM (1998) Molecular determinants of dofetilide block of HERG K$^+$ channels. *Circ Res* 82:386–395.
Frisch MJ, Trucks GW, Schlegel HB, Scuseria GE, Robb MA, Cheeseman JR, and Scalmani G (2009) *Gaussian 09, Revision A.1*, Gaussian, Inc., Wallingford.
Garg V, Sachse FB, and Sanguinetti MC (2012) Tuning of EAG K(+) channel inactivation: molecular determinants of amplification by mutations and a small molecule. *J Gen Physiol* 140:307–324.

Garg V, Stary-Weinzinger A, Sachse F, and Sanguinetti MC (2011) Molecular determinants for activation of human ether-à-go-go-related gene 1 potassium channels by 3-nitro-n-(4-phenoxyphenyl) benzamide. *Mol Pharmacol* 80:630–637.
Gerlach AC, Stoehr SJ, and Castle NA (2010) Pharmacological removal of human ether-à-go-go-related gene potassium channel inactivation by 3-nitro-N-(4-phenoxyphenyl) benzamide (ICA-105574). *Mol Pharmacol* 77:58–68.
Goldin AL (1991) Expression of ion channels by injection of mRNA into *Xenopus* oocytes. *Methods Cell Biol* 36:487–509.
Grunnet M, Abbruzzese J, Sachse FB, and Sanguinetti MC (2011) Molecular determinants of human ether-à-go-go-related gene 1 (hERG1) K+ channel activation by NS1643. *Mol Pharmacol* 79:1–9.
Hansen RS, Diness TG, Christ T, Demnitz J, Ravens U, Olesen SP, and Grunnet M (2006) Activation of human ether-à-go-go-related gene potassium channels by the diphenylurea 1,3-bis-(2-hydroxy-5-trifluoromethyl-phenyl)-urea (NS1643). *Mol Pharmacol* 69:266–277.
Hemmerlein B, Weseloh RM, Mello de Queiroz F, Knötgen H, Sánchez A, Rubio ME, Martin S, Schliephacke T, Jenke M, and Heinz-Joachim-Radzun et al. (2006) Overexpression of Eag1 potassium channels in clinical tumours. *Mol Cancer* 5:41.
Hess B, Bekker H, Berendsen HJC, and Fraaije JGEM (1997) LINCS: A linear constraint solver for molecular simulations. *J Comput Chem* 18:1463–1472.
Hess B, Kutzner C, van der Spoel D, and Lindahl E (2008) GROMACS 4: Algorithms for highly efficient, load-balanced, and scalable molecular simulation. *J Chem Theory Comput* 4:435–447.
Hornak V, Abel R, Okur A, Strockbine B, Roitberg A, and Simmerling C (2006) Comparison of multiple Amber force fields and development of improved protein backbone parameters. *Proteins* 65:712–725.
Hoshi T, Zagotta WN, and Aldrich RW (1990) Biophysical and molecular mechanisms of Shaker potassium channel inactivation. *Science* 250:533–538.
Hoshi T, Zagotta WN, and Aldrich RW (1991) Two types of inactivation in Shaker K$^+$ channels: effects of alterations in the carboxy-terminal region. *Neuron* 7:547–556.
Jiang Y, Lee A, Chen J, Ruta V, Cadene M, Chait BT, and MacKinnon R (2003) X-ray structure of a voltage-dependent K$^+$ channel. *Nature* 423:33–41.
Jones G, Willett P, and Glen RC (1995) Molecular recognition of receptor sites using a genetic algorithm with a description of desolvation. *J Mol Biol* 245:43–53.
Jorgensen WL, Chandrasekhar J, Madura JD, and Impey RW and Klein ML (1983) Comparison of simple potential functions for simulating liquid water. *J Chem Phys* 79:926–935.
Kang J, Chen XL, Wang H, Ji J, Cheng H, Incardona J, Reynolds W, Viviani F, Tabart M, and Rampe D (2005) Discovery of a small molecule activator of the human ether-a-go-go-related gene (HERG) cardiac K$^+$ channel. *Mol Pharmacol* 67:827–836.
Knape K, Linder T, Wolschann P, Beyer A, and Stary-Weinzinger A (2011) In silico analysis of conformational changes induced by mutation of aromatic binding residues: consequences for drug binding in the hERG K$^+$ channel. *PLoS ONE* 6:e28778.
Köpfer DA, Hahn U, Ohmert I, Vriend G, Pongs O, de Groot BL, and Zachariae U (2012) A molecular switch driving inactivation in the cardiac K$^+$ channel HERG. *PLoS ONE* 7:e41023.
Lange W, Geissendörfer J, Schenzer A, Grötzinger J, Seebohm G, Friedrich T, and Schwake M (2009) Refinement of the binding site and mode of action of the anticonvulsant Retigabine on KCNQ K$^+$ channels. *Mol Pharmacol* 75:272–280.
Ludwig J, Terlau H, Wunder F, Brüggemann A, Pardo LA, Marquardt A, Stühmer W, and Pongs O (1994) Functional expression of a rat homologue of the voltage gated either á go-go potassium channel reveals differences in selectivity and activation kinetics between the Drosophila channel and its mammalian counterpart. *EMBO J* 13:4451–4458.
Martin S, Lino de Oliveira C, Mello de Queiroz F, Pardo LA, Stühmer W, and Del Bel E (2008) Eag1 potassium channel immunohistochemistry in the CNS of adult rat and selected regions of human brain. *Neuroscience* 155:833–844.
Mello de Queiroz F, Suarez-Kurtz G, Stühmer W, and Pardo LA (2006) Ether à go-go potassium channel expression in soft tissue sarcoma patients. *Mol Cancer* 5:42.
Nose S (1984) A unified formulation of the constant temperature molecular dynamics methods. *J Chem Phys* 81:511–519.
Pardo LA, del Camino D, Sánchez A, Alves F, Brüggemann A, Beckh S, and Stühmer W (1999) Oncogenic potential of EAG K$^+$ channels. *EMBO J* 18:5540–5547.
Parrinello M and Rahman A (1981) Polymorphic transitions in single crystals: A new molecular dynamics method. *J Appl Phys* 52:7182–7190.
Perry M, Sachse FB, and Sanguinetti MC (2007) Structural basis of action for a human ether-a-go-go-related gene 1 potassium channel activator. *Proc Natl Acad Sci USA* 104:13827–13832.
Sanguinetti MC, Jiang C, Curran ME, and Keating MT (1995) A mechanistic link between an inherited and an acquired cardiac arrhythmia: HERG encodes the I_{Kr} potassium channel. *Cell* 81:299–307.
Sanguinetti MC and Jurkiewicz NK (1990) Two components of cardiac delayed rectifier K+ current. Differential sensitivity to block by class III antiarrhythmic agents. *J Gen Physiol* 96:195–215.
Sanguinetti MC and Tristani-Firouzi M (2006) hERG potassium channels and cardiac arrhythmia. *Nature* 440:463–469.
Schenzer A, Friedrich T, Pusch M, Saftig P, Jentsch TJ, Grötzinger J, and Schwake M (2005) Molecular determinants of KCNQ (Kv7) K$^+$ channel sensitivity to the anticonvulsant retigabine. *J Neurosci* 25:5051–5060.
Schreibmayer W, Lester HA, and Dascal N (1994) Voltage clamping of *Xenopus laevis* oocytes utilizing agarose-cushion electrodes. *Pflugers Arch* 426:453–458.
Seebohm G, Westenskow P, Lang F, and Sanguinetti MC (2005) Mutation of colocalized residues of the pore helix and transmembrane segments S5 and S6 disrupt deactivation and modify inactivation of KCNQ1 K$^+$ channels. *J Physiol* 563:359–369.
Siu SW, Vácha R, Jungwirth P, and Böckmann RA (2008) Biomolecular simulations of membranes: physical properties from different force fields. *J Chem Phys* 128:125103.

Smith PL, Baukrowitz T, and Yellen G (1996) The inward rectification mechanism of the HERG cardiac potassium channel. *Nature* **379**:833–836.

Spector PS, Curran ME, Zou A, Keating MT, and Sanguinetti MC (1996) Fast inactivation causes rectification of the I_{Kr} channel. *J Gen Physiol* **107**:611–619.

Stansfeld PJ, Grottesi A, Sands ZA, Sansom MS, Gedeck P, Gosling M, Cox B, Stanfield PR, Mitcheson JS, and Sutcliffe MJ (2008) Insight into the mechanism of inactivation and pH sensitivity in potassium channels from molecular dynamics simulations. *Biochemistry* **47**:7414–7422.

Stary A, Wacker SJ, Bookharta L, Zacharias U, Karimi-Nejad Y, Aqvist J, Vriend G, and de Groot BL (2010) Toward a consensus model of the HERG potassium channel. *ChemMedChem* **5**:455–467.

Stühmer W (1992) Electrophysiological recording from *Xenopus* oocytes. *Methods Enzymol* **207**:319–339.

Suessbrich H, Schönherr R, Heinemann SH, Lang F, and Busch AE (1997) Specific block of cloned Herg channels by clofilium and its tertiary analog LY97241. *FEBS Lett* **414**:435–438.

Trudeau MC, Warmke JW, Ganetzky B, and Robertson GA (1995) HERG, a human inward rectifier in the voltage-gated potassium channel family. *Science* **269**:92–95.

Warmke J, Drysdale R, and Ganetzky B (1991) A distinct potassium channel polypeptide encoded by the *Drosophila eag* locus. *Science* **252**:1560–1562.

Warmke JW and Ganetzky B (1994) A family of potassium channel genes related to *eag* in *Drosophila* and mammals. *Proc Natl Acad Sci USA* **91**:3438–3442.

Wuttke TV, Seebohm G, Bail S, Maljevic S, and Lerche H (2005) The new anticonvulsant retigabine favors voltage-dependent opening of the Kv7.2 (KCNQ2) channel by binding to its activation gate. *Mol Pharmacol* **67**:1009–1017.

Ye S, Li Y, and Jiang Y (2010) Novel insights into K^+ selectivity from high-resolution structures of an open K^+ channel pore. *Nat Struct Mol Biol* **17**:1019–1023.

Zhang H, Zou B, Yu H, Moretti A, Wang X, Yan W, Babcock JJ, Bellin M, McManus OB, and Tomaselli G et al. (2012) Modulation of hERG potassium channel gating normalizes action potential duration prolonged by dysfunctional KCNQ1 potassium channel. *Proc Natl Acad Sci USA* **109**:11866–11871.

Address correspondence to: Michael C. Sanguinetti, Nora Eccles Harrison Cardiovascular Research & Training Institute, University of Utah, 95 South 2000 East, Salt Lake City, UT 84112. E-mail: sanguinetti@cvrti.utah.edu

I want morebooks!

Buy your books fast and straightforward online - at one of the world's fastest growing online book stores! Environmentally sound due to Print-on-Demand technologies.

Buy your books online at
www.get-morebooks.com

Kaufen Sie Ihre Bücher schnell und unkompliziert online – auf einer der am schnellsten wachsenden Buchhandelsplattformen weltweit! Dank Print-On-Demand umwelt- und ressourcenschonend produziert.

Bücher schneller online kaufen
www.morebooks.de

OmniScriptum Marketing DEU GmbH
Heinrich-Böcking-Str. 6-8
D - 66121 Saarbrücken
Telefax: +49 681 93 81 567-9

info@omniscriptum.com
www.omniscriptum.com

Printed by Books on Demand GmbH, Norderstedt / Germany